MANUAL OF ENTOMOLOGY
AND
PEST MANAGEMENT

LEON G. HIGLEY *Iowa State University*
LAURA L. KARR *Iowa State University*
LARRY P. PEDIGO *Iowa State University*

MACMILLAN PUBLISHING COMPANY
NEW YORK
COLLIER MACMILLAN PUBLISHERS
LONDON

Research Assistance by Heath Petersen
Selected illustrations by Karen Burkwall Johnson

SB
931
.H44
1989

Macmillan Publishing Company
866 Third Avenue, New York, New York 10022

Collier Macmillan Canada, Inc.

ISBN 0-02-393350-X

Printing: 1 2 3 4 5 6 7 8   Year: 9 0 1 2 3 4 5 6 7 8

PREFACE

This book provides practical information on insect pest management and is designed as a text for laboratories in insect pest management and as a supplement to lecture courses having a pest-by-commodity orientation. The manual does not require other reference material and can stand alone as a class reference. However, the book was written as a practical complement to **Entomology and Pest Management** by Larry Pedigo, which emphasizes the theory and principles of insect pest management, and both books will be most valuable when used together. When the same concept is addressed in both books, we have used different examples and approaches in the manual to avoid repetition and improve comprehension. Neither book presumes any previous background in entomology or pest management.

FORMAT

Chapters follow a common format. Each chapter begins with a written discussion of the chapter topic, including definitions of new terms (terms are boldfaced in the text). After the discussion, a list of objectives is presented along with a list of terms, suggested displays or exercises, study and discussion questions, and an annotated bibliography. The bibliography and accompanying notes are not intended to be exhaustive but rather reflect the authors' preferences and opinions. A few chapters also include additional information in appendices to the chapter. Space is provided for changes and additions by an instructor to allow for modification. Chapters are divided into sections on basic entomology, groups of insect pests (such as agronomic, medical, and urban pests), and aspects of management including sampling, decision making, and control tactics.

Although sufficient information is included in each chapter for self-study, the manual is more useful when used in conjunction with laboratory instruction. There is no substitute for direct experience examining living and preserved insects. Each chapter roughly corresponds to material for a laboratory session, although a number of chapters would need more than one period, and other chapters would need much less than a period. In courses requiring a previous entomology course, the first four chapters on basic entomology can be omitted or used for review.

CONTENT

The content and emphasis of this book reflect both our opinions and philosophies regarding entomology and pest management. We have not discussed in depth many of the broader, theoretical topics that are the basis for insect pest management, because these are treated in **Entomology and Pest Management**. However, we have addressed concepts and issues that are central to practical pest management. Among the themes we have developed across chapters is the concept of tolerating injury as an approach to understanding how and why pest management differs across commodities. Additionally, we have used the keystone concept of the economic injury level as another mechanism for characterizing these differences. We have avoided simply listing details on identifying individual pest species or control practices, because we believe the specifics on insect identification, life history, and management practices are most appropriately provided through an instructor to reflect circumstances in a state or region. Although some bias towards the Midwest may be reflected in this work, we have attempted to address insect pest management for the entire US and Canada.

Frequently, insect pest management laboratories focus on memorizing identification and life history information for dozens of pests. We have tried to de-emphasize pest identifications (which are easily forgotten by students) and focus instead on procedures for identifying insects and evaluating pest problems, which we feel is of more lasting value. The insect pests we have listed are those we recognize as most important, but obviously other entomologists may have different choices. Important commodities were identified through US agricultural statistics, and insect pest species of these commodities were chosen based on opinion and objective indications of importance from references, if available. Current reference information often overlooks pests of certain commodities, such as range grasses and mushrooms, and we made a particular effort to include these pests in our lists.

The management chapters describe the array of management options available, but we recognize that students may have little practical association with certain procedures, such as genetic control. We particularly emphasized insecticidal control including calibration and safety, out of the realization that this is one, potentially hazardous, procedure students will almost certainly encounter. The chapter on regulatory management reinforces another theme we have carried through the book - the importance of introduced pests.

In applied entomolgy, so much attention and effort is directed at identifying, understanding, and solving insect pest problems that it is easy to loose sight of the biological importance and fascination of insects. One virtue of the insect pest management approach is that it emphasizes the ecological importance of insects. Our hope is that students using this manual not only learn how to practice insect pest management, but also come to appreciate insects as fascinating and desirable parts of the natural environment.

ACKNOWLEDGEMENTS

Many people have aided us in completing this book and deserve our thanks. We are especially indebted to Heath Petersen who assisted with research and provided many useful suggestions and insights. His help was crucial on many chapters. We also thank Karen Burkwall Johnson for her excellent work on illustrations for Chapter 2. Page layout, formating, and camera-ready copy were prepared by LGH using WordStar Professional 5.0 software, Hewlett Packard TimesRoman and Helvetica soft fonts, and an HP LaserJet Series II printer.

A number of our colleagues at Iowa State University have reviewed chapters, made suggestions, offered expert advise, and otherwise helped in preparing the book. These colleagues include: (in the Department of Entomology) William Berry, Joel Coats, Aaron Gabriel, John Obrycki, Wayne Rowley, and Wendy Wintersteen; and (in the Department of Plant Pathology) Phyllis Higley and Laura Sweets. We also are grateful for reviews of the entire manuscript by Robert Snetsinger, Department of Entomology, Pennsylvania State University, and David Hogg, Department of Entomology, University of Wisconsin-Madison. However, the authors are solely responsible for any errors.

Finally, we appreciate the efforts of Macmillian Publishing Company in producing this book, and particularly the assistance of Beth Anderson and Bob Rogers.

CONTENTS

Section I. Basic Entomology   1

Chapter 1. Principles of Insect Identification and Preservation   3

Chapter 2. Insect Morphology   16

Chapter 3. Insect Systematic   56

Chapter 4. Insect Growth and Development   108

Section II. Insect Pests   125

Chapter 5. Agronomic Pests   128

Chapter 6. Horticultural Pests   139

Chapter 7. Forest, Shade Tree, and Related Pests   156

Chapter 8. Medical and Veterinary Pests   167

Chapter 9. Stored Product Pests   182

Chapter 10. Urban Pests   191

Section III. Management   203

Chapter 11. Sampling and Decision Making   204

Chapter 12. Insecticidal Management   218

Chapter 13. Biological and Genetic Control   254

Chapter 14. Ecological Management and Host Plant Resistance   263

Chapter 15. Regulatory Management   273

SECTION I   BASIC ENTOMOLOGY

The first four chapters provide fundamental information on insect biology that is essential for insect pest management. Specific topics in these chapters are principles of insect identification and preservation, insect morphology, insect growth and development, and insect systematics. Many of these subjects are necessary in learning how to identify insects, the first of the four steps in curing an insect pest problem (these steps are identifying the insect, quantifying the numbers or effects of the insect, evaluating if management is warranted, and undertaking the appropriate management tactic). However, understanding basic aspects of entomology is equally important in recognizing how insects become pests and how aspects of insect life history can be exploited for pest management.

These basic chapters only are brief treatments of broad subjects, but they still include considerable information. In particular, learning the fundamentals of insect morphology and insect systematics requires understanding many new terms and recognizing many features of the insect body. Mastering this terminology is essential for working in insect pest management. Much of the material you need to know is most easily learned by working with preserved and living insects. Examining insects is crucial in learning insect morphology and in learning how to sight identify insects to order.

Perhaps the greatest challenge in learning basic entomology is in the tremendous variety of insect forms and life histories. Although the diversity in insect shapes and habits can be an impediment to understanding entomology, many commonalities are recognizable across insect groups. Moreover, many entomologists are attracted to the field because of this diversity. If you can cultivate an interest in insects through studying basic entomology, you will better understand insect pest management as well as better appreciate insects as a part of the natural environment.

GENERAL BIBLIOGRAPHY

Borror, D. J., D. M. DeLong, and C. A. Triplehorn. 1981. An introduction to the study of insects. 5th ed. Saunders College Pub., NY
- includes information on sampling, preservation, and extensive keys to insect families. A valuable reference for North American insects.

Atkins, M. D. 1978. Insects in perspective. Macmillan Pub. Co., Inc., New York, NY
Evans, H. E., ed. 1984. Insect biology. Addison-Wesley Pub. Co., Reading, MA
- many introductory entomology texts are available, and these two are among our favorites. Both are well-illustrated and easy to read.

Evans, H. E. 1978. Life on a little-known planet. E. P. Dutton, New York, NY
- originally published in 1968, this is an eloquent, non-technical, discussion of insects. Well-written, witty, informative, and enjoyable.

O'Toole, C. 1985. Insects in camera. Oxford Univ. Press, New York, NY
    - a "photographic essay on behavior" that includes superb photographs by Ken Preston-Mafham illustrating aspects of insect life history.

Pedigo, L. P. 1989. Entomology and pest management. Macmillan Pub. Co., New York, NY
    - the companion text to this manual.

1 PRINCIPLES OF INSECT IDENTIFICATION AND PRESERVATION

Insect identification (the placing of individual insects into established classification groups) is of fundamental importance to pest management and to entomology in general. We cannot properly recognize or evaluate an insect problem without first identifying the insect causing the problem. Although different insects may produce comparable injuries on a plant, our methods for sampling, for evaluating the problem, and for insect control may differ, depending on the specific insect involved. Proper identification is even more important for entomologists studying insects. Lack of identification makes research on an insect impossible; incorrect identification makes research worthless, and worse, an impediment to science.

One example that illustrates these points relates to the role of insects in the transmission of the yellow fever virus. Around 1880, Carlos Finlay, a Cuban physician, formulated and tested a theory that yellow fever was transmitted by mosquitoes. This was an idea ahead of its time. Unfortunately, in conducting his experiments Finlay did not use a single species of mosquito, but had colonies of mixed species. Because only certain mosquito species can transmit the yellow fever pathogen, Finlay's results were variable and inconclusive. Not until 1900, in work by Walter Reed and other members of the United States Yellow Fever Commission, was mosquito transmission of yellow fever proven. Because mosquito species were not identified correctly in Finlay's research, twenty years were lost.

Just as recognizing different species was crucial in the yellow fever story, so is the concept of species central to our discussion of insect identification. When we speak of identification, we may refer to many different levels of classification: phylum (Arthropoda - arthropods), class (Insecta - insects), order (Coleoptera - beetles), or family (Curculionidae - weevils). However, these higher levels of classification are in large part artificial; they are based on our perceptions of relationships between species. In fact, our only certain level of classification is the species itself.

SPECIES

A **species** is a group of interbreeding organisms that can produce offspring with the capacity for viable reproduction. A horse and donkey can mate and produce offspring (mules), but they are not of the same species because mules cannot reproduce. Although we can conduct mating tests between individuals to determine if they are of the same species, as a practical matter this is rarely done. More often, morphological features are used to distinguish between species. In some groups of organisms the species concept does not work as well as in others; for example, our definition has limited applicability to viruses. Likewise, the use of morphological features to distinguish species also has limitations; for example, many fungi are so dissimilar in their life stages that different forms may have different species names. Fortunately, the situation is not so confused with most insect species.

An insect species is designated by a formal **species description** that provides a species name and sufficient information to differentiate that species from all others. A species description is based on a single specimen, called a **type**, that is the model or basis of the species. If a question arises, for example about whether or not a given species is unique, it will be answered by comparing the type of the species to types of other species. Type specimens are of tremendous scientific value; most are deposited at major collections such as the British Museum or the US National Museum (Smithsonian Institute). When types are destroyed, as happened in Germany during World War II, our ability to differentiate or relate species is seriously weakened. Species descriptions and types are the foundations for all insect identification.

Nomenclature

One way to look at identification of a species is to ask the question "What is the unique name of this organism?" Although Juliet may argue "What's in a name? That which we call a rose by any other word would smell as sweet," scientists have a much more conservative view of **nomenclature** (the naming of biological units). Indeed, animals are given scientific names based on formal, legalistic procedures called "The Rules of Zoological Nomenclature" (botanical nomenclature, naming plants, follows different rules). You need to understand the naming of insects, because insect names (unfortunately) can change.

Scientific Names

All recognized species of animals and plants are given a scientific name which is binomial (consisting of two words). The first word is the genus name (with a capitalized first letter) and the second word is the specific name (which is never capitalized). Usually, the name of the individual who described the species is written after the scientific name the first time it is used in an article. If the author's name is given in parentheses, this indicates that the species is in a different genus than when it was originally described. Remember we said that levels of classification above the species are rather artificial, they rely on interpretation and best opinion. Often a species thought to be in one genus will later be placed in a different genus because new relationships are uncovered. Sometimes examination of type specimens will show that individuals described as two species are actually one, resulting in a change of name for one of the species. Scientific names of insects, including economically important insects, may undergo a surprising number of changes. For example, the seedcorn maggot, now called *Delia platura*, was first described in 1826 as *Anthomyia platura* and has been known by 33 different names. These **synonymies** (different names for the same species) can be important in pest management if you need to access scientific articles about a species.

Common Names

Unlike mammals and birds, most insects do not have common names. However, virtually all economically important species, as well as other insects of general interest (such as most butterflies), are given formal common names by the Entomological Society of America

(ESA).   The ESA publishes lists of these common names and rejects or recommends new common names as they are proposed. This formal procedure helps eliminate different species having the same common name or one species having many common names. (Multiple common names for a few species are so well established that they are still formally recognized; for example, *Heliothis zea* is known by three common names: corn earworm, tomato fruitworm, and bollworm. Additionally, two common names sometimes occur for different stages of the same species; e.g., scarab beetle larvae are called white grubs, but the adults are called June beetles.) Most often common names are not longer than three words. A common name usually includes two parts: one part indicating the group to which the insect belongs and another part which modifies the group name (for example, house fly, bed bug, cat flea, or Colorado potato beetle). If the group name is not scientifically accurate, it is combined with the modifier. For instance, citrus whitefly (order Homoptera) is not a true fly, whereas Mexican fruit fly is a true fly (order Diptera). Modifiers include host (e.g., onion maggot), range or origin (e.g., southwestern corn borer, European corn borer), or a physical description (e.g., oystershell scale, fourlined plant bug).

Distinguishing Species

The identification of insects to species can pose problems. Sometimes an insect species turns out to be a group or complex of many species. Occasionally, insects in two species are actually members of the same species. Such situations usually develop because the characteristics used to distinguish between species are obscure. Indeed, sometimes it is impossible to distinguish between species in some immature stages or for a given sex. Unlike birds or mammals, the majority of insect species can only be identified by an expert. Moreover, many insect species, particularly in the tropics, are unknown to science and have never been described.

For insect pest management we need to identify insects to the species level. Fortunately, most economically important species are so well known that we do not need to rely on experts or refer to species descriptions. In this, and subsequent laboratories, you will learn procedures and methods for identifying insects. Although memorization is important for learning individual species, your ability to remember species and to identify those you have never seen before will be enhanced by learning principles of insect identification.

IDENTIFICATION

The vast number of insect species makes it impossible for us to recognize insects as, for instance, we might recognize birds. Entomologists can recognize insects to order and usually to family but beyond these levels it is often hit and miss (probably as many misses as hits). The variety in insect forms both within and between species, the number of species, the small size of insects, and the frequent obscurity of references for specific insect groups all challenge and impede our identifications.

Insect identification relies on a knowledge of **insect morphology** (the forms and structures of insects) and **insect classification** (the ordering of insects into groups on the basis of their relationships). Learning insect morphology and classification is like the question of the chicken or the egg; we need to know classification to understand how forms and structures of insects vary, but we need to know morphology to understand how insects are arranged into groups. Just as you might expect, the tremendous numbers and diversity of insects make insect morphology a broad topic. For example, one glossary of entomology, which consists mostly of morphological terms, has over 12,000 entries! Likewise, insect classification covers a vast territory. Insect classification includes two areas, **taxonomy** (the theoretical basis for classifying organisms), which we've already touched on, and **systematics** (the study of the kinds and diversity of organisms and the relationships among them). A knowledge of systematics is crucial in identification.

To identify an insect the first objective is to recognize the order it to which it belongs. We will discuss insect systematics in Chapter 3, but you should appreciate the importance of knowing orders and higher classifications. Although we may recognize a number of insect species by sight, invariably most we encounter will be unknown to us. Knowing the order provides a useful starting point for more detailed identification and also provides some knowledge about the natural history of the insect. The farther down the classification scheme we identify an insect, the more we will know about it. If we know an insect is a beetle, for instance, then we know it has immature stages that do not resemble the adult (specifically, it has larval and pupal stages) and that it probably has one (or just a few) generations per year. If we know it is in the family Curculionidae (a weevil), then we know that the insect is a plant feeder, whose immatures probably feed inside plant tissues and whose adults feed outside tissues by drilling holes in plant parts. You must learn to identify insects to order by sight.

Once an insect is identified to order, how do we learn the species? For most insect species we would need to move down the classification hierarchy to determine the family, then the genus, and finally the species. However, most economically important species are sufficiently well-known that we don't have to move down this ladder. Usually, once we know the order, we can try to determine an economic insect to species. But a few words of warning. Not all insects you will want to identify will be economically important species. Sometimes you will have a specimen and you will want to know if this insect is a pest or a nonpest. If it is a pest species, you may know it, or information will be available for its identification. But if it is not a pest, perhaps the best you'll be able to say is what it is not, not what it is. Although you will learn principles of identification, how to sight identify any insect to order, and how to identify many economically important species, you should realize that this is far short of the training necessary for you to identify most insects to species.

An identification can be based on three types of evidence: (1) **morphology**, form and structures, (2) **products or effects** of insects, or (3) **situation**, time of occurrence, location, or host of the insect. Depending on the specific insect and circumstances, any one type of evidence may be sufficient for an identification. However, definitive identification of an insect will depend on morphological characters, and consequently, morphology is emphasized in identification. Nevertheless, indices and situations of insects provide valuable support for morphological identifications, as well as being useful in their own right. Your ability to recognize insects and insect problems will be greatly enhanced by understanding the natural history (behavior, ecology, etc.) of pest species.

Morphology

Morphology can be used in two ways to identify an insect: through recognizing individual structures or through gestalt. Most identifications are accomplished by examining individual morphological features. We look for characters or modifications of characters that are unique for the species. Often a single character is sufficient to describe a species; for example, the black dagger shape on a black cutworm moth wing, the toothed margin of saw-toothed grain beetle, the clear hind wing and red abdomen of a squash vine borer, or the red hourglass design on the abdomen of a black widow spider. However, many identifications require multiple characters. A common method of using multiple characters is through analytical keys. An **analytical key** is a devise for identifying an organism, in which paired choices are presented, followed by additional choices that ultimately lead to the correct identification. These choices may involve features such as presence or absence, size or shape of a structure. (Appendix 3.1 is an example of an analytical key, a key to insect orders.) The usefulness of analytical keys is dependent on how well we can differentiate between characters. Obviously, using analytical keys will require at least a basic understanding of insect morphology. Probably few entomologists are familiar with all the terminology and structures associated with all groups of insects. However, all entomologists do recognize basic morphological maps for major insect groups. By using **landmark features**, morphological structures that serve as reference points for other structures, we can learn and recognize more specific characters as necessary (e.g., Fig. 1.1).

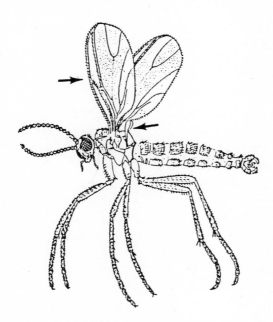

Figure 1.1   A true fly, Order Diptera, Family Sciaridae. Landmark features include one pair of forewings and a pair of halteres (knobbed structures). (Courtesy Pennsylvania State University)

Gestalt refers to the integration of structures and form of an organism that is not discernible by considering individual parts. Many authorities on insect groups can immediately recognize various species, but cannot explain the precise characters they used to separate the species. In fact, it is the sum of many morphological characteristics that allow for the identification. We use this idea of gestalt, or the entirety of characters, when we identify insects by pictures, with **pictorial keys**, or by comparison with preserved insects. (Pictorial keys may also focus on individual morphological features, as do analytical keys [e.g., Fig. 1.2].) This approach is appropriate for many insects; for example, identification of North American butterfly species, even subspecies, can be accomplished by looking at the overall wing pattern. Generally, the more familiar you become with a species, the less you will need to rely on individual characters to recognize the species. But some groups or species may always require detailed examination and the use of keys.

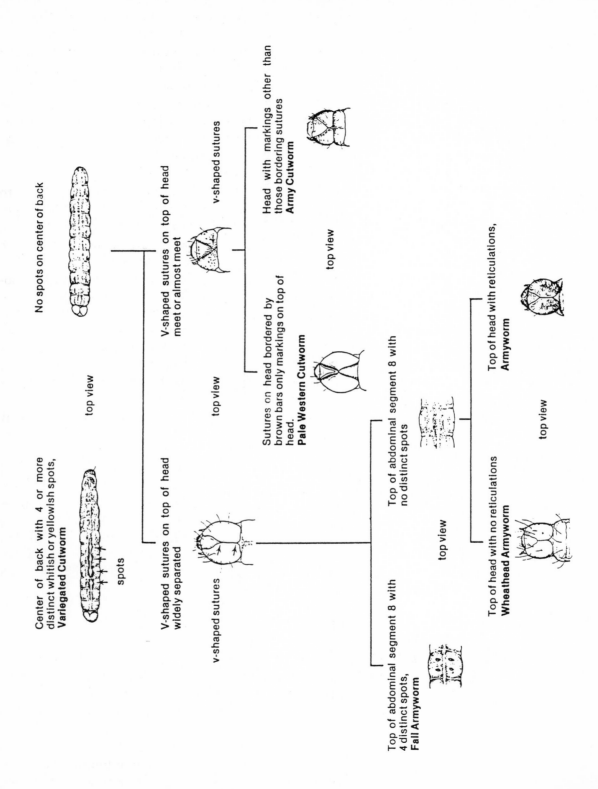

Figure 1.2   Pictorial key to caterpillars on wheat in the northcentral US.   (Courtesy Kansas State University)

Morphological features can be useful for more than species identification. Frequently, we need to know the sex or the age of an insect. Most insects display some **sexual dimorphism**, differences in body form or function based on sex. Besides the obvious difference in genitalia, which may not be very apparent in some groups, other features that commonly differ between sexes include body size, antennal type, and arrangement of the eyes. Estimates of age for adult insects are most often used in research to distinguish between different generations. One common procedure is to look at ovarian development in females as an indication of age. For insect pest management, we often are interested in the age of immature insects. Sometimes body size is used to estimate the stage of development, but body size can be misleading. One frequently used measure for immatures is the size of the head capsule, which is more readily associated with a given stage than is a certain body size.

Morphological evidence can present some challenges for identification, especially to the unwary. In particular, **divergence**, differences in structure between closely related groups, and **convergence**, similarities in structure between unrelated groups, may cause confusion. Divergence among closely related species is common among insects. For example, the golden tortoise beetle, bean leaf beetle, and Colorado potato beetle are all leaf beetles (family Chrysomelidae) whose immatures display considerable divergence in body form, having a dorsoventrally compressed spiny body, a narrow, elongate body with hardened anterior and posterior regions, and a swollen, rounded body, respectively. Divergence also occurs within species, for instance, many insects have immature stages that are completely dissimilar to the adult stage. Within a stage you may have **polymorphism**, differences in body forms and structures. We've mentioned sexual dimorphism, which is one type of polymorphism, but other types of variability are possible. Size differences, variations in coloring or in patterns, and winged or wingless conditions are some common manifestations of polymorphism.

Convergence probably fools more neophyte entomologists than does divergence. Many insects have developed similar structures, yet are unrelated. For example, the true flies (order Diptera) can be distinguished from other groups because flies have only a single pair of wings; however, many other groups of insects have mechanisms for linking the fore and hind wings together, so they function and resemble a single wing. **Mimicry**, the resemblance of one organism to another, unrelated organism, is common among insects; for example, many innocuous insects have color patterns like wasps and bees. Often, looking for landmark features or a key structure will reveal the true identity of a mimic. Indeed, by focusing on the unique features of a group (be it order, family, or species) and ignoring more superficial similarities or differences, you can avoid identification problems with divergence and convergence.

Products or Effects and Situation

The other sources of evidence for identification, products or effects and situation, often are used in combination. Products or effects include all signs left by insects (such as webbing or silk, cast skins, and excrement [called **frass**]) and symptoms of insect injury on plants or animals . Sometimes these indications may be specific to a species, although more often they merely indicate the presence of an insect. Situation encompasses many factors such as geographic location, location on a plant, time of the year or in the life cycle of the host, and type of host. Knowing the host plant or animal of an insect is especially valuable in limiting the possibilities. Products or effect and situation also provide a check on morphological identifications. The more thorough your knowledge of insect pests and their biology, the less chance you have of making gross mistakes in identification.

PRESERVATION

Insect preservation is extremely important for scientific purposes but less so for pest management. For scientific research, often preserved insect samples, called **voucher specimens**, are labeled and placed in university or museum collections to permit review if the identity if the insects being investigated is ever in question. For insect pest management, new or very uncommon pests need to be preserved to allow expert examination. Consequently, we need to know how to properly preserve material for those instances when we are unable to identify an insect in the field. Poor or incorrect preservation of insects makes identification difficult or impossible. Additionally, you may need to send insects to extension entomologists or other specialists for identification, and your specimens will need to be properly preserved. In this and subsequent laboratories you will work with many preserved insects. For preserved insects to be of value they must be undamaged, therefore, treat all specimens you handle as if they are type specimens.

In the laboratory on sampling you will learn various techniques for collecting and quantifying insects. Once you have unknown insects, how do you store them? Insects can be stored for years in 70-80% ethanol, although moth and butterfly scales, fine hairs, and similar structures may be lost or obscured. Refrigerating or freezing insects frequently is acceptable for short term storage. Dry storage of insects usually is not acceptable, because as the insects dry out, body parts easily break off. Additionally, dried material must be relaxed, or remoistened, before it can be mounted.

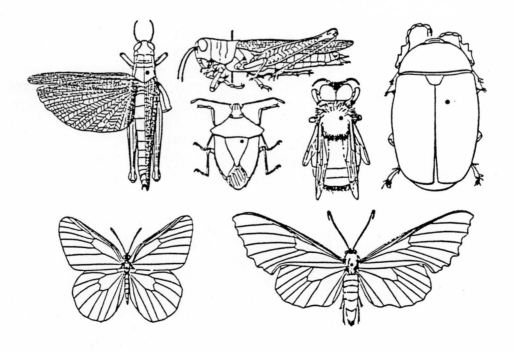

Figure 1.3 Correct pinning location for various insects. (Courtesy USDA)

Mounting Insects

Mounting involves the presentation of insects to allow convenient handling of the insect without damaging the specimen and to clearly display all features of the insect necessary for identification. The rule of thumb for mounting is that hard-bodied material is placed on pins, and all soft-bodied material is placed in alcohol or another liquid preservative. Thus, spiders, ticks, mites, and insect larvae are all stored in liquid (or on microscope slides). Proper preservation of insect larvae is a challenge because colors are easily lost in the preservative. Putting larvae into boiling water for a few minutes denatures proteins which helps limit color loss once the specimen is placed in alcohol.

Insects are mounted on pins either by **direct pinning**, in which a pin is inserted through the insect's body, or by **double mounting**, in which an insect is pinned with a minuten (a very small pin) or glued to a card point and then attached to an insect pin. Insects should only be mounted on proper insect pins. Insect pins are made from stainless steel (to prevent rust) and usually are lacquered (japanned) so insects adhere better to the pin. Pins are available in various sizes (diameters); sizes 2 and 3 are useful for most insects. Insects should be fresh or relaxed before pinning to avoid breaking off legs, antennae, or other structures. Insects are pinned vertically through the body, just right of the midline (so that one side of the insect is visible) (Fig. 1.3 and 1.4). The height of the insect on the pin will vary depending on the size of the insect, but it should be high enough to avoid damage to legs during handling. Pinning blocks, that have steps of various heights, can be used to arrange insects or labels on pins.

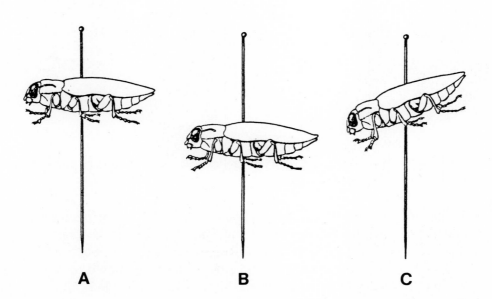

A   **B**   **C**

Figure 1.4   Insect pinning. A. Correct pinning orientation and height. B. Incorrect orientation. C. Incorrect height (too low). (Courtesy USDA)

Double mounts are used for small insects (Fig. 1.5). An insect should be double mounted if placing it directly on a pin would obscure features. A common technique in double mounting is to glue insects to a cardboard point on an insect pin. However, points are inappropriate for some types of insects, for example, small moths. In these instances, insects are pinned on minuten pins, which are then placed on cork or pith blocks on a regular insect pin.

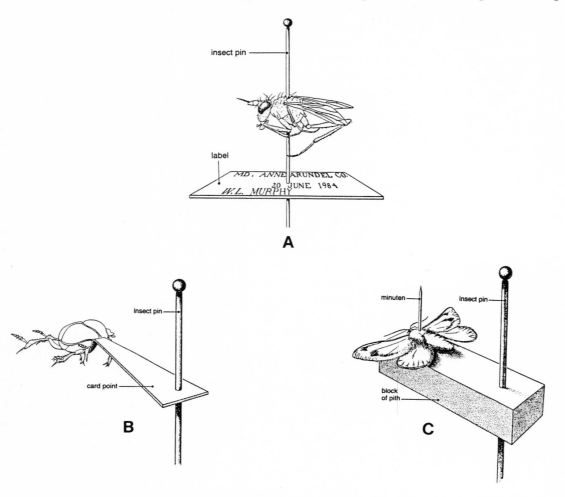

Figure 1.5 Insect mounts. A. Direct mounting on pin with proper label. B. Double mount on cardboard point. C. Double mount on minuten pin with pith. (Courtesy USDA)

All preserved material must be labeled; unlabeled material is worthless scientifically. A label is placed on the pin (in the the vial for material in alcohol) and includes location, date, collector, and host (if available). The more detailed a description of location the better, although the detail may be limited by space on the label. The date should include day, month, and year. If information is available on the host, this should be included (on a separate label on the pin if necessary).

Lepidoptera (moths and butterflies) must have their wings spread after pinning. **Spreading** involves placing the wings out at right angles to the body so that the fore and hind wings are visible. A spreading board is used to position the wings and to allow them to dry at the appropriate angle (Fig. 1.6). Considerable care is required in spreading to avoid tearing the wings or dislodging wing scales.

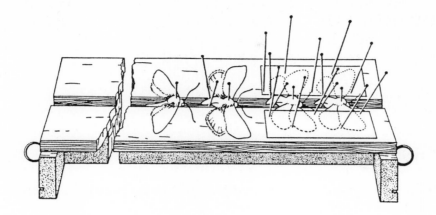

Figure 1.6 Spreading board with moths. (Courtesy USDA)

Shipping Insects

Occasionally, you may need to mail insects for identification. Extension entomologists routinely receive insects mailed in pill bottles, in cardboard boxes, or even stuck on pieces of tape in envelopes. Identifying insects in envelopes after they have gone through a franking machine can challenge the most knowledgeable expert. The two concerns in shipping insects are loss and damage of specimens. Simple precautions such as including address information inside the package and using certified mail for important material can minimize the possibility of loss. Preserved material in vials should be tightly stoppered and each vial wrapped separately - glass should not touch glass. The chief risk with pinned specimens is that the pins will come loose during shipment. Pins should be placed firmly into a pinning bottom, with sufficient space around each specimen for easy removal. A piece of cardboard should be placed on top of the pins with cotton or foam between the cardboard and the lid of the box, so that pins are held firmly in place. Vials or pinned material should be placed in a small box that is surrounded by packing material in a larger box. There are no restrictions on shipping dead insects within the US or outside the country. However, restrictions do apply to living insects, so all insects sent for identification should be dead.

OBJECTIVES

1) Understand the species concept and how insect species are designated including species descriptions, types, and nomenclature.

2) Understand the roles of insect morphology and classification in identification.

3) Recognize the three types of evidence for insect identification (morphology, productsi or effects, and situation), and know the uses and limitations of morphological evidence.

4) Be able to properly preserve insects and to prepare specimens for shipment.

5) Additional objectives (at instructor's discretion):

OPTIONAL DISPLAYS AND EXERCISES

1) Identify the specimens that are divergent or convergent in the presented groups:
   a. divergent groups
   b. bee or wasp mimics.

2) Properly pin and label the insects provided.

3) Prepare a box of specimens for shipment. Your instructor will test your packing.

4) Examine the presented insect signs and insects associated with these indices.

5) Additional displays and exercises (at instructor's discretion):

TERMS

analytical key	gestalt	pictorial keys	spreading
classification	identification	polymorphism	synonymies
convergence	landmark features	products or effects	systematics
direct pinning	mimicry	sexual dimorphism	taxonomy
divergence	morphology	situation	type
double mounting	mounting	species	voucher specimen
frass	nomenclature	species description	

DISCUSSION AND STUDY QUESTIONS

1) If two different species of insects injure plants in precisely the same fashion, why do we need to differentiate between these species (or do we need to differentiate)?

2) How is insect identification similar to and different from identifying other organisms, or

even other objects (cars, aircraft, etc.)?

3) What are the primary objectives in preserving insects?

4) How can you determine if different insect specimens are in the same species?

5) Describe how a species is formally designated.

6) What is the importance of insect classification to pest management?

7) What factors do you use to identify common insects (such as mosquitoes, beetles, butterflies, cockroaches)?

BIBLIOGRAPHY

Pedigo, L. P. 1989. Entomology and pest management. Macmillan Pub. Co., New York, NY
- Chapter 3. Insect classification.

Borror, D. J., D. M. Delong, and C. A. Triplehorn. 1981. An introduction to the study of insects, 5th ed. Saunders College Pub., New York, NY
- includes information on sampling, preservation, and extensive keys to family. This book is an essential reference for anyone working with North American insects.

Mayr, E. 1969. Principles of systematic zoology. McGraw-Hill Book Co., New York, NY
Simpson, G. G. 1961. Principles of animal taxonomy. Columbia Univ. Press, New York, NY
- both of these books provide excellent discussions on the theory and practice of taxonomy and systematics in zoology. They are thorough and technical yet surprisingly readable.

Steyskal, G. C., W. L. Murphy, and E. M. Hoover. 1986. Insects and mites: techniques for collection and preservation. USDA Misc. Pub. No. 1443
- a comprehensive discussion of methods for collecting and preserving insects. Includes keys to order and ordinal descriptions.

2 INSECT MORPHOLOGY

Insect morphology is fundamental to any study of entomology. Indeed, the many subdisciplines of entomology all rely on morphology to at least some degree. As we discussed in the previous chapter, morphology is a vital complement to systematics. Similarly, understanding insect morphology is crucial in many aspects of insect behavior and ecology. And because insects are not well represented in the fossil record, our comprehension of insect evolution primarily depends on comparative morphology. For pest management, insect morphology is most important as it relates to systematics and insect identification. But other aspects of morphology relating to areas such as ecology or evolution can be of practical importance. For example, comparative morphology may indicate evolutionary relationships between pests; closely related pest species may have similar or common natural enemies for use in biological control.

Nevertheless, the overriding importance of morphology to pest management comes from the need to understand morphology for insect identification. In the last chapter we discussed some aspects of morphology as they pertain to identification. Obviously, we need to be able to recognize major insect structures, particularly those that may serve as landmark features. We also need to be familiar with structures that may be less obvious but that are frequently important for identification. Additionally, the notions of convergence and divergence are extremely important for an understanding of morphology. In fact, looking at divergence in structures will dominate much of our study of morphology.

STUDYING INSECT MORPHOLOGY

Although insect morphology is a broad topic, we do not need to learn all insect structures and modifications. Instead, we need to become familiar with a relatively small set of basic structures that are common to most insects. These basic structures will be modified, often radically, in different groups of insects, so we also need to recognize the most common of these modifications. In considering the diversity of insect features two types of relationships among structures are important. **Homologous structures** are features that have a common evolutionary origin, but may or may not have similar functions. All insect wings are homologous; they have a common origin. Structures that have similar functions but different origins are **analogous structures**. Insect wings and bird wings are analogous. We will observe many examples of homology and analogy in examining modifications of insect structures. For example, although insect legs may be greatly modified with differing functions among species, all insect legs are homologous. On the other hand, many insect species have sucking mouthparts with similar functions and appearance but that are actually modifications of different structures and, therefore, analogous (e.g., Fig. 2.10, 2.11, and 2.12). In addition to changes in form, structures may be lost or become greatly reduced, possibly losing functionality. Such trace or rudimentary structures are called **vestigial**.

Before we begin considering morphology in more detail, it is appropriate to have something of a pep talk at the outset. Because understanding insect morphology requires that

you learn many new terms and recognize an array of insect structures, memorization is necessary. Blindly learning structures and names is likely to become a brain-numbingly boring task.
To avoid such a perspective, you need to approach the material with an inquisitive outlook and
realize that morphology involves more than just learning structures and names. The excite-
ment and fascination entomologists have with morphology is comparable to that which you
may get from reading a mystery novel - both present puzzles and provide clues to answering
those puzzles. Often morphological questions have no ready answer. And we have limited
certainty that any answers we develop are correct. However, asking and investigating morphological questions provide new insights and understanding. Moreover, as we pose questions
when studying morphology, we not only increase our interest in the material but also have the
opportunity to better appreciate the relationship between morphology and other aspects of
entomology.

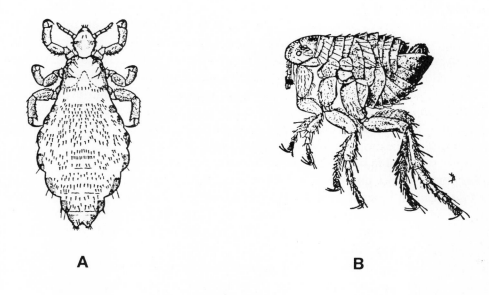

A   **B**

Figure 2. 1   Two human parasites. A. The head louse, *Pediculus humanus capitis*, which is dorsoventrally compressed (from top to bottom). B. The human flea, *Pulex irritans*, which is laterally compressed (from side to side).
(A courtesy USDA; B courtesy USDHEW-CDC)

One example of a morphological question we might ask is why most insect parasites of
vertebrates are dorsoventrally compressed (flattened from top to bottom) whereas fleas are
laterally compressed (flattened from side to side) (Fig. 2.1). Think about it for a moment.
Many parasites probably are flattened so they can maintain closer contact with their host and
consequently are harder to dislodge. If that's true, shouldn't fleas be the same? We are all
familiar with the ability of fleas to jump, so we know that fleas may not stay with their host all
the time. In fact, most fleas lay their eggs off of the host, so fleas need to be mobile. Being
laterally compressed makes it easier for fleas to move between host hairs and get on and off
the host. Thus, many parasites have a body form to allow them to stick on their host while
fleas have a body form that allows movement from the host. This explanation (which probably
isn't the whole story) illustrates how considering morphological differences may emphasize or

help explain differences in life history between species.

MORPHOLOGY AND EVOLUTION

Another point you should understand about structures and their function, is that evolution does not produce ideal, perfect structures. Instead, a given feature is modified through evolution in response to changing needs. The insect wing base and folding mechanism is an example of this modification process. Unlike birds and bats whose wings are modifications of a foreleg, insect wings are not derived from an appendage. In birds and bats, leg joints became wing joints, but such joints are lacking for the insect wing. To design a structure that would allow rotation, flexing, and folding of a wing, we might consider a ball and socket joint or analogous structure as an ideal. Such a joint is vastly different from the structures that evolved, because evolution occurs only through variation and selection of existing structures (in this case the existing structure is the integument, or hardened skeleton, of the insect body). Figure 2.6 illustrates an insect wing base which, combined with associated muscles, permits a variety of wing movements and wing folding. That this combination of sclerites (hardened plates), membranes, and muscles allows such an array of movement is remarkable, a testament to evolutionary trial and error. It illustrates that natural structures do not form in pursuit of a theoretical ideal, rather they represent workable solutions arising through evolutionary alteration and modification. Possibly more than any other group, insects provide endless examples of such adaptations. Thus, in studying insect morphology we are faced with challenging diversity but also with wondrous examples of evolutionary engineering.

Morphology includes both structure and function, but our emphasis will be on structures, in other words, insect anatomy. Nevertheless, you need to understand the function of major insect features. In looking at dead, preserved insects, it is easy to overlook how insects perform intricate actions, movements, and functions. Consequently, one valuable method for learning how insect parts operate is to watch living insects feed, fly, walk, oviposit, mate, and conduct the other activities of life.

This introduction to insect morphology includes three sections. First, we will consider the terminology used to describe insect parts. Next we will examine the major insect structures in a relatively unmodified insect, the grasshopper. Finally, we will look at common modifications of structures in other insect groups. The following sections focus on these areas and include: a written exercise for terminology of orientation, directions for examination and dissection of the lubber grasshopper, and descriptions and comparisons of variations in structures.

OBJECTIVES

1) Understand the terminology of orientation indicated in Table 2.1.

2) Be able to recognize the basic morphological features of insects. In particular, know the components of the mouthparts, antennae, and legs.

3) Know the structures and vocabulary listed in the terms section.

4) Recognize common morphological modifications including head orientation, antennal types, and leg types.

5) Additional objectives (at instructor's discretion):

LIST OF DISPLAYS AND EXERCISES

1) Terminology of orientation. Learn the descriptive terminology of morphology and complete a written exercise. Instructions start on page 26.

2) Basic insect morphology: examination of the lubber grasshopper. Learn the external and internal features of a generalized insect by dissecting the lubber grasshopper. Instructions start on page 30.

3) Morphological modifications. Recognize common morphological modifications exhibited in displayed insects. Instructions start on page 44.

OPTIONAL DISPLAYS AND EXERCISES

1) Dissect provided insects and compare to lubber grasshopper to become familiar with the diversity in internal features.

2) Examine displayed insects illustrating extreme morphological modifications.

3) Observe the displayed living insects or video tapes to appreciate how insect structures function in life.

4) Additional displays or exercises (at instructor's discretion):

TERMS

alate
analogous
    structures
antennae
apterous
brachypterous
cerci
cervix
clypeus
common oviduct
compound eyes
coxa
crop
cuticle
dorsal blood
    vessel
elytra
epipharynx

esophagus
exoskeleton
fat body
femur
flagellum
foregut
hemocoel
hemolymph
hindgut
homologous
    structures
hypopharynx
integument
intestine
labella
labial palpi
labium

labrum
lateral oviducts
Malpighian tubules
mandibles
maxillae
maxillary palpi
mesothorax
metathorax
midgut
ocelli
ovaries
ovipositor
pedicel
pharynx
prothorax
pterous
rectum

scape
sclerites
sclerotized
spermatheca
spiracles
sutures
tarsomeres
tarsus
tentorium
testes
tibia
tracheae
trochanter
vas deferens
ventral nerve cord
ventriculus
vestigial

DISCUSSION AND STUDY QUESTIONS

1) Genitalic features demonstrate tremendous diversity in form among insects. Why do these structures show so much variability?

2) Which morphological features seem to be most consistent in form across different insect groups? Why?

3) List insect features or modifications that indicate the life history or habits of an insect species.

4) The exoskeleton protects internal organs and functions as an armored covering; however, insects also may be flexible and can achieve a vast array of movements. How are both protection and movement possible?

5) How is the division of the insect body into head, thorax, and abdomen reflected in functional differences between these regions?

BIBLIOGRAPHY

Pedigo, L. P. 1989. Entomology and pest management. Macmillan Pub. Co., New York, NY
    - Chapter 3. Insect structures and life processes.

Most introductory texts include sections on morphology. Additionally, many guides or books on specific insect groups may have information on morphological features important for identification.

Chapman, R. F. 1982. The insects. 3rd ed. Harvard Univ. Press, Cambridge, MA
- a valuable treatment of insect structures and function, with emphasis on insect physiology.

Fox, R. M. and J. W. Fox. 1964. Introduction to comparative entomology. Reinhold Pub. Co., New York, NY
- although dated, an introductory text noteworthy for its considerable emphasis on comparative morphology. A readable volume with clear illustrations.

Jones, J. C. 1981. The anatomy of the grasshopper (*Romalea microptera*). Charles C. Thomas Publisher, Springfield, IL
- a comprehensive guide to the structures of the lubber grasshopper. It provides a particularly thorough treatment of the musculature.

Snodgrass, R. E. 1935. Principles of insect morphology. McGraw-Hill Book Co., New York, NY
- this is the classic reference on insect morphology. Many sections are showing their age, and Chapman (cited above) does offer some more current evaluations. Nevertheless, no reference equals Snodgrass in scope, information, or usefulness. Although reprinted for decades, Snodgrass is (unhappily) out of print.

Torre-Bueno, J. R. de la. 1978. A glossary of entomology. New York Entomological Society, New York, NY
- a comprehensive dictionary to entomology. Rather dated (the 1978 edition is a reprint of the 1937 original), but still useful. Both Torre-Bueno and Snodgrass are vital entomological references that desperately need updating or new replacements.

FIGURE CITATIONS

Atkins, M. D. 1978. Insects in perspective. Macmillan Pub. Co., Inc., New York, NY

Borror, D. J., D. M. DeLong, and C. A. Triplehorn. 1976. Introduction to the study of insects. 4th ed. Holt, Rinehart, and Wilson, New York, NY

Jones, J. C. 1981. The anatomy of the grasshopper (*Romalea microptera*). Charles C. Thomas Publisher, Springfield, IL

Matheson, R. 1948. A laboratory guide in entomology for introductory course. Cornell University, Ithaca, NY

McAlpine, J. F. (ed.) 1981 and 1987. Manual of nearctic Diptera. Vol 1 and 2. Agriculture Canada, Research Branch, Ottawa, Ont

Pfadt, R. E., editor. 1985. Fundamentals of applied entomology. 4th ed. Macmillan Pub. Co., New York, NY

Snodgrass, R. E. 1935. Principles of insect morphology. McGraw-Hill Book Co., New York, NY

Snodgrass, R. E. 1946. The anatomy of the honeybee. *in* The hive and the honeybee. Dadant and Sons, ed. Dadant and Son, Hamilton, IL

Torre-Bueno, J. R. de la. 1978. A glossary of entomology. New York Entomological Society, New York, NY

West, L. S. 1951. The housefly. Comstock Pub. Co., Inc., Ithaca, NY

Wilson, M. C., D. B. Broersma, A. V. Provonsha. 1984. Practical insect pest management. Vol. 1. Fundamentals of applied entomology, 2nd ed. Waveland Press, Inc., Prospect Heights, IL

DISPLAYS AND EXERCISES

Exercise I. Terminology of Orientation

To learn insect morphology we must be able to correctly recognize new structures from written descriptions and be able to correctly refer to structures ourselves. Accurately describing or referring to individual features depends on having a precise terminology for various directions or surfaces. Table 2.1 summarizes terms used to characterize locations or orientations of insect parts. Many terms you are already familiar with, although others may be less common. Notice that the form of the term differs depending on how it is used (as an adjective, adverb, or noun). Adjective forms usually end in -al and mean "of or on a location" (e.g., lateral = of or on the side); adverb forms end in -ly (most can also end in -ad rather than -ly, e.g., laterad) and mean "towards a location" (e.g., laterally = towards the side); noun forms usually end in -on or -um and mean a surface (e.g., dorsum = the top surface). Adjective descriptors for parts of segments refer to a segmental surface rather than a direction (e.g., dorsal = of or on the top, whereas notal = of or on the notum or top surface).

Some locations or oblique directions are described by combining different words. In particular, prefixes are used to indicate oblique directions by combining with other words (e.g., dorsolateral). These terms are not exclusive, in fact a single direction or location may be properly described by many different terms. For example, the location of a structure on the left side of the head can be called anterolateral, laterocephalic, or anterodextrolateral (quite a mouthful). Although many constructions may provide proper descriptions of a location or direction, you will notice in your readings that certain descriptors or combinations are more commonly used than are others. Learn to understand this terminology so you can recognize locations and structures in your readings. Similarly, you should be able to use these terms so you can describe structures you identify.

Table 2.1   Directions and descriptors for orientation.

Domain	Adjective	Adverb	Noun	Prefix	Location
General:					
	anterior	anteriorly	anterior	antero-	front
	posterior	posteriorly	posterior	postero-	rear
	lateral	laterally	--------	latero-	side
	medial	medially	medial	--------	middle
	median	medianly	median	--------	middle
Appendages:					
	apical	apically	apex	--------	tip
	distal	distally	--------	--------	away from body
	proximal	proximally	--------	--------	near body
Body Regions or Parts:					
	cephalic	cephalically	cephalon	cephalo-	head
	caudal	caudally	cauda	caudo-	anal end
	dextral	dextrally	dextron	dextro-	right
	sinistral	sinistrally	sinistron	dextro-sinistro	left
	mesal	mesally	meson	meso-	center
	dorsal	dorsally	dorsum	dorso-	top
	ventral	ventrally	venter	ventro-	bottom
	ental	entally	enton	ento-	internal
	ectal	ectally	ecton	ecto-	external
Body Segment:					
	notal	--------	notum	noto-	top surface
	pleural	--------	pleuron	pleuro-	side surface
	sternal	--------	sternum	sterno-	bottom surface
	tergal	--------	tergum	tergo-	top surface

Written Exercise: Describing Orientation

For the accompanying illustrations indicate the proper descriptor for each surface (indicated by a capital letter) and each arrow direction (indicated by a number). More than one term or combination of terms may be correct for each surface or direction.

Surfaces (noun form):

A _____

B _____

C _____

D _____

E _____

F _____

Directions (adjective form):

1 _____

2 _____

3 _____

4 _____

5 _____

6 _____

7 _____

8 _____

9 _____

10 _____

11 _____

12 _____

13 _____

14 _____

15 _____

16 _____

17 _____

18 _____

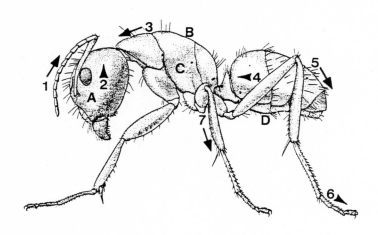

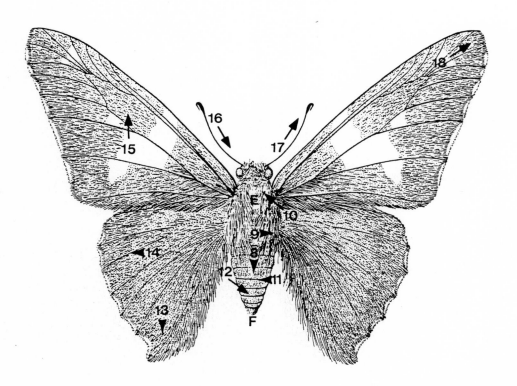

Exercise I.   Figures for describing orientation. Ant (Hymenoptera) and butterfly (Lepidoptera). (Courtesy USDA)

Exercise II. Basic Insect Morphology:

Examination of the Lubber Grasshopper

A working knowledge of insect morphology requires that you learn various insect structures and their locations. Because grasshoppers display the basic morphological structures without substantial modifications, they are useful in learning the fundamentals of insect morphology. For your study you will examine the lubber grasshopper, *Romalea microptera*, which is one of the largest grasshoppers in North America. We will learn structures on the adult grasshopper, but most of these will be the same or similar in grasshopper nymphs. However, in other groups the internal and external morphology can be radically different between adults and immatures, as you will see in Chapter 3.

Read through the descriptions and discussion before working with a grasshopper. Follow the written descriptions and illustrations by locating features on your specimen. Note in particular the relationship of one part to another and the form, features, adjacent structures, etc. that define a given structure. Drawing structures may be helpful in learning some parts. Once you can recognize structures on the lubber grasshopper, you will be prepared to learn how these basic features are altered and rearranged in other insect groups.

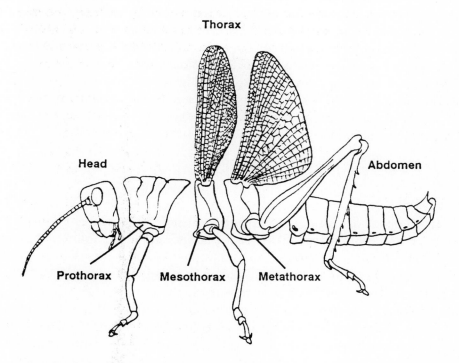

Figure 2.2   Major body regions of the lubber grasshopper, *Romalea microptera*. (Modified by KBJ after USDA)

External Morphology

Obtain a grasshopper for examination. Figure 2.2 illustrates major anatomical features. The insect body is divided into three parts: **head**, **thorax**, and **abdomen**. The head appears as a single structure, although it is thought to be derived from the fusion of five or six segments. The thorax is divided into three distinct sections: **prothorax**, **mesothorax**, and **metathorax**. A pair of legs is associated with each thoracic section and a pair of wings is located on both the meso- and metathorax. The abdomen is composed of ten segments. Other insects may differ in the number of abdominal segments, but 11 or 12 segments are thought to occur primitively. Very often segments are fused together or lost, so it is difficult to establish a definitive number.

The outer covering of the insect is called the **integument** or **cuticle**. In total these body walls are known as the **exoskeleton**. The exoskeleton functions as an armored covering for the insect and also provides attachments for muscles. Besides surrounding and protecting internal organs, the exoskeleton also may form invaginations, or inner folds. The most extreme example of such folds occurs in the insect head, where cephalic invaginations resembling arms are fused together to form the **tentorium**, which provides internal support and strength (Fig. 2.3C). However, other than the tentorium, insects generally lack internal strengthening, with virtually all support and skeletal functions being provided by the exoskeleton.

Examine the integument of the head, thorax, and abdomen in more detail. Notice that depending on location the integument differs in color, hardness, and flexibility. The head is greatly hardened to protect the sense organs and brain, as well as to support the mouthparts. The thorax is only slightly less hardened, because it must support the legs, wings, and their associated muscles. Generally, the abdomen is least hardened and most flexible. Notice that the integument is formed of hardened plates, called **sclerites**, with lines between sclerites called **sutures**. Sutures are important features in demarcating one sclerite or region from another. Sometimes sclerites are separated by membranous regions. Lift and gently pull apart two adjacent abdominal tergites (dorsal sclerites). Notice that these are joined by a heavy membrane. Membranous regions allow the expansion and contraction of the abdomen, which may be important for oviposition, when the abdomen becomes greatly elongated to allow egg laying into the soil. For some insects having an expandable abdomen is vital to accommodate food, such as when a mosquito is engorged with blood.

The Head and Mouthparts

Use a hand lens or dissecting microscope to examine the head. Figure 2.3 illustrates features of the head and mouthparts. The head is connected to the thorax by a membranous region called the **cervix**, or neck (Fig. 2.3E). The head capsule, excluding the cervix and mouthparts is the **cranium**. The top of the cranium is called the **vertex** (Fig. 2.3A, B, C, and E). Near the vertex are two large, multifaceted **compound eyes**. Two small, simple eyes called **ocelli** (singular **ocellus**) are adjacent to the compound eyes, as are the bases of the antennae. That region of the head between the antennae and below the vertex is the **frons**. The sides, or cheeks of the face are called **genae** (singular **gena**). The **frontal suture**, which extends from the bottom of the eye to the base of the clypeus, distinguishes the gena from other head regions.

Examine the mouthparts (Fig. 2.3D-I). At the lower front part of the head the structure with a straight upper margin and irregular sides and lower margin is the **clypeus** (Fig. 2.3F). Hinged to the lower margin of the clypeus is the **labrum**. Lift up the labrum and note its size and shape. On the underside of the labrum is a lobe, the **epipharynx** (pronounced epy-fair-inks), that functions in taste (Fig. 2.3D). The labrum and epipharynx often are thought of as a single structure, the labrum-epipharynx.

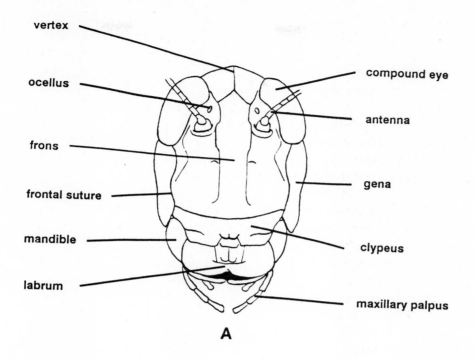

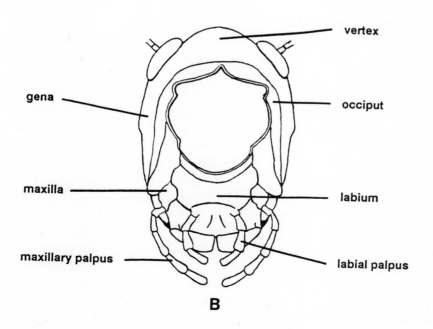

Figure 2.3 Aspects of the lubber grasshopper head. A. Anterior view. B. Posterior view. (Modified by KBJ after Matheson 1948)

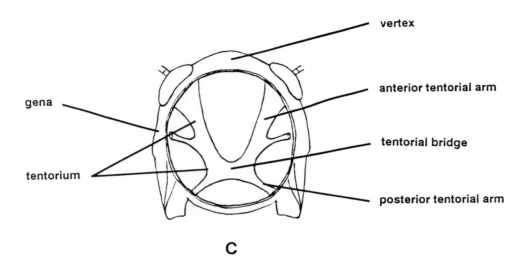

C

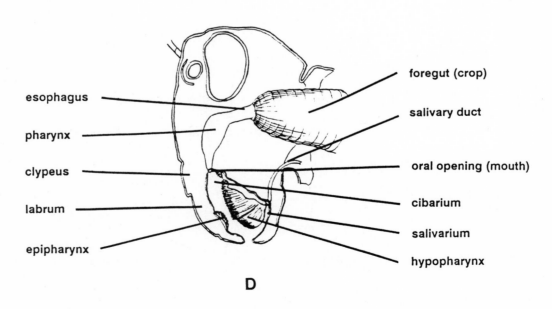

D

Figure 2.3   Aspects of the lubber grasshopper head, continued. C. Posterior internal view, indicating the tentorium. D. Lateral internal view. (C KBJ; D modified by KBJ after Matheson 1948)

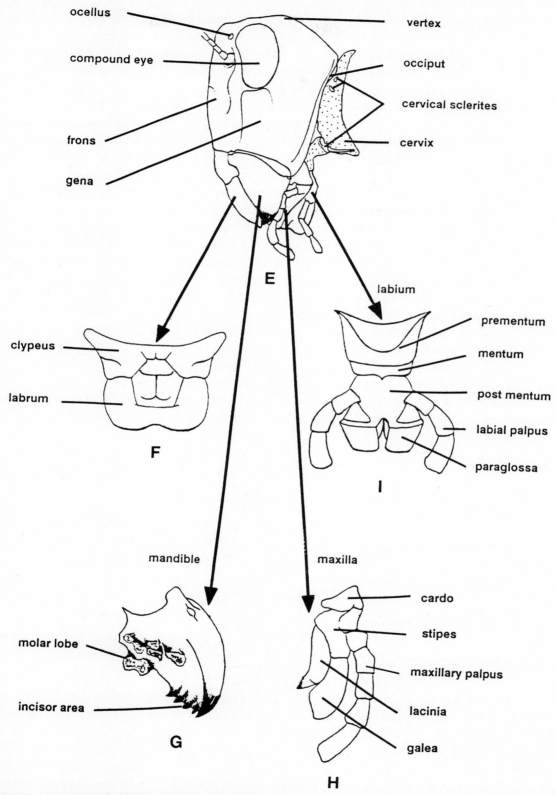

Figure 2.3   Aspects of the lubber grasshopper head, continued. E. Lateral view. F. Detail of clypeus and labrum. G. Detail of mandible. H. Detail of maxilla. I. Detail of labium. (Modified by KBJ after Matheson 1948)

Remove the labrum and clypeus and observe the arrangement of mouthparts. The **mandibles** are large, triangular structures that attach and articulate with the head at the base of the genae. The mandibles are very heavily **sclerotized**, or hardened, because they are used in tearing and chewing food. Remove the mandibles, and examine the incisor area on the mandibles (Fig. 2.3G). Notice the well-developed muscles inside the mandibles that attach to the cranium. These muscles are necessary to provide strong tearing and chewing movements.

Behind the mandibles are the **maxillae** (singular **maxilla**). The maxillae and the labium behind the maxillae both function in manipulating the food. Remove the maxillae and examine one of them in more detail (Fig. 2.3H). A maxilla is composed of five parts: dorsally is the **cardo**, beneath it the **stipes**, laterally the **galea** and a five segmented **maxillary palpus**, and below the stipes the **lacinia**.

Behind the maxillae is the **labium**. Between the maxillae and in front of the labium is the lobe-shaped **hypopharynx**, which functions somewhat like a tongue in tasting and manipulating food (Fig. 2.3D). The open region in front of the hypopharynx and behind the clypeus forms a preoral cavity called the **cibarium**. The true mouth or oral opening is at the top of the cibarium. The small pocket behind the hypopharynx and in front of the labium is the **salivarium**, where the salivary duct opens. In other insects the salivarium is greatly modified. Remove the labium for detailed examination (Fig. 2.3I). The labium is composed of the dorsal, u-shaped **prementum**, with the **mentum** and **postmentum** located ventrally (these parts may not be easily distinguished). Three segmented **labial palpi** occur laterally, and below the postmentum are the lobe-shaped **paraglossa**.

Learn to recognize these components of the mouthparts and their relationship to each other. Grasshopper mouthparts are thought to be similar to those of primitive insects, but other insect groups demonstrate tremendous modifications. A thorough understanding of mouthparts in the grasshopper will assist you in recognizing mouthpart modifications in other groups.

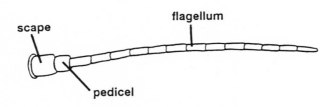

Figure 2.4 Insect antenna. (KBJ)

Next we will consider the **antennae** (Fig. 2.4). An insect antenna has three parts: a proximal segment called the **scape**, another segment called the **pedicel**, and a distal segment called the **flagellum**. The flagellum is divided into numerous subsegments or annulations. How do we distinguish between true segments and subsegments? True segments will have attached muscles that extend to the next segment. However, as a practical matter entomologists often ignore this distinction. For example, many insect keys will use characteristics such as 10 antennal segments; what this really means is that the insect has a scape, pedicel, and 8 subsegments of the flagellum. The easiest way to recognize the three parts of the antennae is to count segments distally from the antennal base. The antennae primarily function as sensory organs and are highly variable among insect groups, although all antennae retain the three basic parts.

The Thorax

Identify the three segments of the thorax (Fig. 2.2). Notice that in the lubber grasshopper the pronotum (the notum of the prothorax) is saddle-shaped and extends caudally to cover most of the mesonotum. Examine the notum, sternum, and pleuron for each thoracic segment. To support the wings and legs and to allow for muscle attachment, the thoracic sclerites are tightly fused together. Try to recognize individual sclerites by looking for sutures. Laterally, on the meso and metathorax are breathing pores called **spiracles**. A pair of spiracles can occur on any thoracic or abdominal segment, however, in most species spiracles are not present on every segment. The spiracles have associated musculature to allow opening and closing.

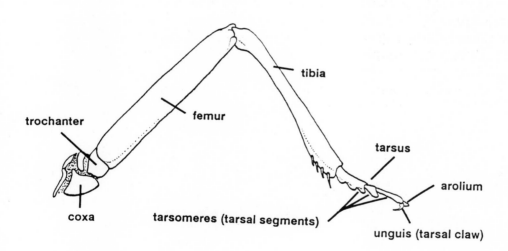

Figure 2.5   Prothoracic leg of the lubber grasshopper. (KBJ)

A pair of legs is located on each of the thoracic segments. The legs may differ somewhat in size and shape between segments, but all insect legs share common features. Examine a prothoracic leg. The leg is composed of five segments (Fig. 2.5). The proximal segment is the **coxa**. In some species the coxa may be difficult to distinguish from thoracic sclerites. Adjacent to the coxa is the **trochanter**, followed by the **femur** and **tibia**. Compare the femur of the prothoracic leg to that of the metathoracic leg. Notice that the metathoracic femur is much more developed and enlarged for jumping. The final leg segment is the **tarsus**. Just as the flagellae of the antennae are divided into subsegments, so are tarsi divided into subsegments.

These subsegments are called **tarsomeres** or **tarsal segments**, although these are not true segments in the sense of having individual muscles. Primitively the tarsus was a single segment, and a few insect groups (orders Collembola and Protura) retain this trait. However, ancestors to the winged insects apparently had a modified tarsus with five tarsomeres, and now most insects have anywhere from two to five tarsomeres. In fact, the lubber grasshopper has a modified tarsus in which the first three tarsomeres are fused together, so that only three tarsal segments are evident. Additional features may occur on the most distal tarsomere and are represented on the lubber grasshopper. One set of structures is a pair of claws, or **ungues** (singular **unguis**). Between the ungues is a second structure, a fleshy pad or lobe called the **arolium**. Some insects have small pads beneath the ungues called **pulvilli** (singular **pulvillus**) or pads beneath each tarsomere that are called **tarsal pulvilli**. Identify the leg parts on the pro-, meso-, and metathoracic legs of your specimen.

The other important appendages associated with the thorax are the wings. The number, size, shape, venation, and texture of wings vary among insects. Some groups are primitively wingless, whereas others have secondarily lost their wings. Those insects with wings are called **alate** or **pterous**, and those without wings are called **apterous**. Insects with greatly shortened wings are called **brachypterous**.

Examine the fore and hind wings of your specimen. Grasshoppers have a modified fore wing, called a **tegmen**, which is hardened and leathery. The hind wing is larger and membranous. Usually, insect wings are membranous with long, sclerotized wing veins, however, wing modifications, some radical, are common. The wing veins help strengthen and support the wing. Squares or boxes formed by veins are called cells, and their size or location, in addition to that of the veins, can be useful in identification.

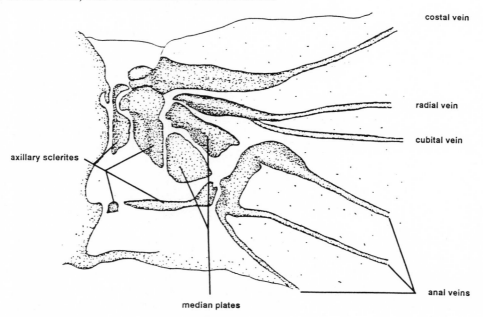

Figure 2.6   Fore wing base of the lubber grasshopper. (LGH)

Examine the base of the fore wing (Fig 2.6). Notice how the wing veins originate, and how membranes and sclerites are arranged. These sclerites associated with the wing base are called **axillary sclerites**. The axillary sclerites are part of structures that permit wing movement, including flexing and folding.

The Abdomen

Examine the abdomen of your specimen. Externally, the abdomen seems relatively unmodified. On the pleural region of the first abdominal segment is an oval, membrane covered spot called the **tympanum** (plural **tympana**), which is an auditory receptor. Not all insects have tympana. In other groups the tympanum may be located on the tibia, the metathorax, or even the wing base. Pairs of spiracles occur on the first eight abdominal segments.

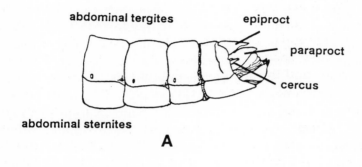

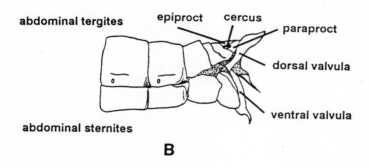

Figure 2.7   Terminal abdominal segments of the lubber grasshopper. A. Male. B. Female. (Modified by KBJ after Matheson 1948)

The terminal abdominal segments show much greater modification than the preceding segments (Fig. 2.7). In the female, the four horn-like structures make up the **ovipositor**, or egg-laying organ. The median, dorsal sclerite is called the **epiproct** and the lateral, dorsal sclerites the **paraprocts**. Between the epiproct and paraproct are short, horn-shaped **cerci** (singular **cercus**). Features of the terminal abdominal segments and external genitalia (copulatory organs) of insects are tremendously variable. These structures can be complex, difficult to recognize and may have a unique terminology. An extreme example is in male fleas whose genitalia are so complicated that their method of operation is unknown for most species. Insect genitalia can be extremely valuable in identifying some insects to species.

Internal Morphology

Although many internal systems are similar in most insect species (e.g., the nervous and circulatory systems), others are quite variable (e.g., the digestive system). Figure 2.8 presents assorted lateral views of different internal systems. You may find it helpful to make sketches of these systems from a dorsal aspect.   You should know the components of various internal systems and be able to recognize these features in a specimen.

Begin your examination by removing the head. With a scalpel or scissors, cut around and through the cervix. The large tube that is severed is the anterior end of the crop or the posterior end of the esophagus. Using scissors or a strong blade, cut medially from the posterior of the vertex to the frons, trying not to cut into tissue beneath the exoskeleton. Pull apart the head and pin the open sections in a dissecting tray to allow for examination of internal features. (Because the head is so greatly hardened it may be difficult to dissect the head to reveal internal structures; consequently, besides those cuts indicated, you may need to make additional incisions.) Notice the sclerotized arms forming the tentorium. Follow the **esophagus** to the **pharynx** and the mouth (Fig. 2.8A). Also examine (Fig. 2.8B) the **brain** and identify the **optic lobes** and the **subesophageal ganglion**. With magnification and a careful dissection, you should be able to see nerves extending from the ocelli and antennae to the brain.

For the remaining structures, you need to dissect the thorax and abdomen. Remove the legs and wings. You may find it easiest to cut the cuticle with fine scissors. Make a median, dorsal cut from the thorax to the terminal abdominal segments. Again, try to avoid cutting tissue beneath the cuticle. Once the incision is complete, use dissecting pins to spread apart the integument and pin open the specimen in the dissecting tray.

The **dorsal blood vessel**, or heart, lies close to the upper surface of the abdomen (Fig. 2.8D). Small openings, or **ostia**, allow entry of the blood, or **hemolymph**, and the **alary muscles** pump the blood anteriorly (the alary muscles may not be apparent). An upper cavity, the **pericardial sinus**, with the dorsal blood vessel lies above the **dorsal diaphragm**. Beneath the dorsal diaphragm is the major body cavity, the **perivisceral sinus**, followed by the **ventral diaphragm** which separates the **perineural sinus**. The entire body cavity is termed the **hemocoel**.

Cut through the dorsal diaphragm (if it is still intact). The silvery tubes running through the hemocoel are **tracheae** (singular **trachea**), or air tubes (Fig. 2.8C). The tracheae extend through the body cavity from the spiracles. The large thoracic air sac is an enlargement of the tracheal system which is necessary to meet the high oxygen demands of leg and wing muscles.

Examine the digestive system (Fig. 2.8A). The insect gut can be divided into the **foregut**, which includes the **pharynx, esophagus, crop** (a food storage organ), and **proventriculus** (which is not well defined in the lubber grasshopper), the **midgut**, which consists of the ventriculus and is the primary organ for digestion and assimilation, and the **hindgut**, which includes the **intestine** and **rectum**. The foregut is separated from the midgut by the **cardiac valve** and the midgut from the hindgut by the **pyloric valve**. Additional structures associated with the digestive system are the **gastric caeca**, which are blind outpockets of the anterior portion of the midgut, and the **Malpighian tubules**, important excretory organs that open into the anterior end of the hindgut. Try to identify the major parts of the digestive system. Besides these organs, the hemocoel may include a loose network of small, white tissue, the **fat body** which have cells that store food but also conduct important metabolic activities (e.g., carbohydrate, amino acid, and protein metabolism).

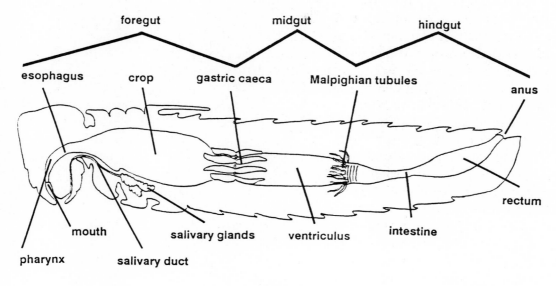

foregut   midgut   hindgut

esophagus   crop   gastric caeca   Malpighian tubules   anus

mouth   salivary glands   ventriculus   intestine   rectum

pharynx   salivary duct

A

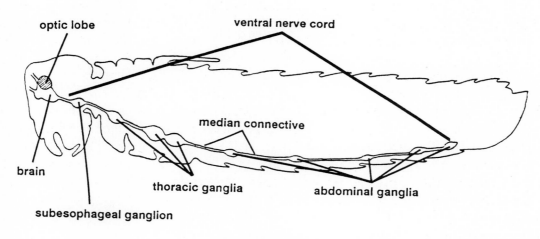

optic lobe   ventral nerve cord

median connective

brain   thoracic ganglia   abdominal ganglia

subesophageal ganglion

B

Figure 2.8   Lateral schematic views of internal systems in the lubber grasshopper. A. Digestive system. B. Nervous system. (Modified by KBJ: A after Jones 1981; B after Pfadt 1985)

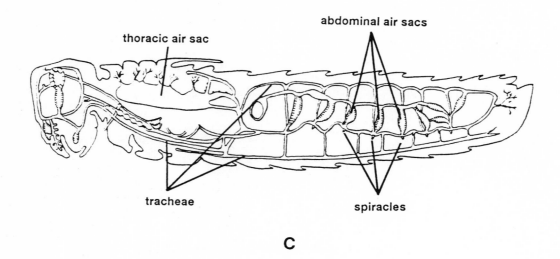

C

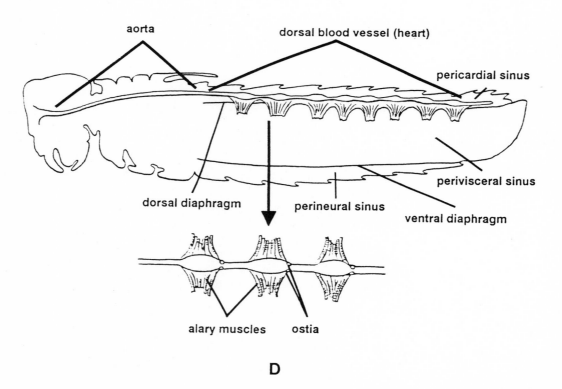

D

Figure 2.8   Lateral schematic views of internal systems in the lubber grasshopper, continued.   C. Respiratory system.   D. Circulatory system.   (Modified by KBJ: C after Matheson 1948; D after Jones 1981)

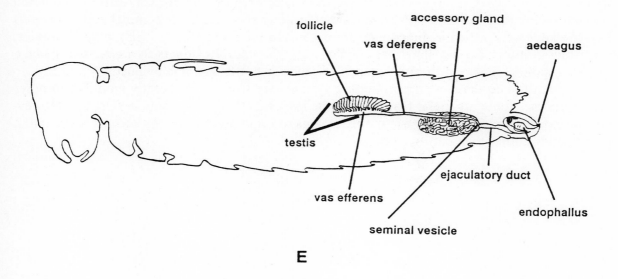

E

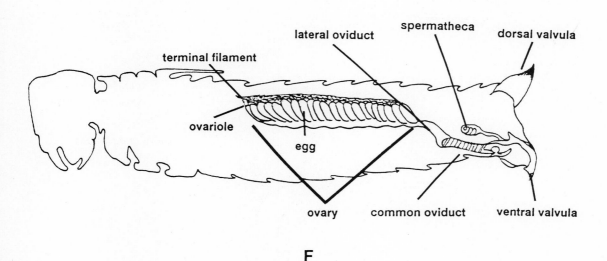

F

Figure 2.8   Lateral schematic views of internal systems in the lubber grasshopper, continued. E. Male reproductive system. F. Female reproductive system. (Modified by KBJ after Jones 1981)

You have already examined the anterior portion of the nervous system (Fig 2.8B). Posteriorly, the **ventral nerve cord** lies below the ventral diaphragm and includes various thoracic and abdominal ganglia.

The male reproductive system (Fig. 2.8E) consists of paired **testes** (singular **testis**), which are comprised of numerous **follicles**. The **vas deferens** connects each testis to a common **seminal vesicle**, and an **accessory gland** also opens into this duct. The seminal vesicle extends into the aedeagus, which is the copulatory organ. In the female (Fig. 2.8F), paired **ovaries**, consisting of individual **ovarioles**, are connected by the **lateral oviducts** to a **common oviduct**. The **spermatheca**, a sperm storage organ, also connects to the common oviduct.

Both the digestive and reproductive systems are likely to be greatly modified in other insect groups. Some organs or features may be lost and others more fully developed. However, these differences primarily are modifications of fundamental organs such as occur in the grasshopper, rather than entirely new structures.

Exercise III. Morphological Modifications

Morphological features vary among species of any group of organisms, but insects are remarkable for the degree to which differences occur. Insects vary in form, size, shape, color, and virtually any other characteristic. Structures can be so reduced or modified as to become almost unrecognizable. However, morphological modifications do not occur at random. Features change to accommodate different functions, and the diversity of insect structures is a reflection of the diversity of their habits.

In this exercise you will review some common modifications in morphological features. Try to recognize how the generalized features you examined in the lubber grasshopper are altered in other insect groups. Additionally, many of the modifications you will consider illustrate differences in life histories between insect species. Some modifications may reflect specialized habits, whereas others may allow more diverse actions. Use the figures as a guide to identifying individual features.

Head

Many modifications of the head are possible. In fact, in some insect larvae a recognizable head does not occur. One type of modification you should recognize is in the orientation of head and mouthparts. Figure 2.9 indicates the three common types of mouthpart orientation: **hypognathus** (mouthparts directed downward), **prognathus** (mouthparts directed forward), and **opisthognathus** (mouthparts directed backward).

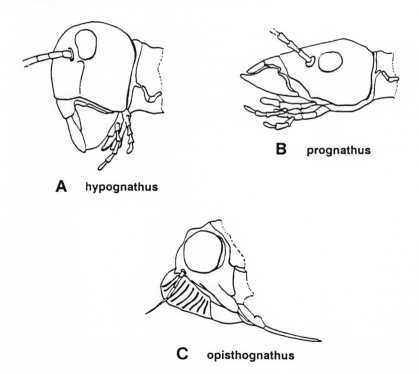

Figure 2.9   Common types of mouthpart orientation. A. Hypognathus (directed downward). B. Prognathus (directed forward). C. Opisthognathus (directed backward). (Modified by KBJ after Atkins 1978)

Mouthparts

The lubber grasshopper has generalized chewing mouthparts, which are thought to be similar in form to primitive insect mouthparts. However, many insect groups have mouthparts that are modified for lapping, sponging, or sucking. These differences in feeding apparatus allow insects to ingest a variety of liquid, semi-solid, and solid foods. In some groups adult mouthparts are vestigial or lost, and no feeding occurs in the adult stage. Figures 2.10-2.14 illustrate some of the major forms of mouthpart modifications that occur in different insect orders.

In the orders Hemiptera and Homoptera (Fig. 2.10), the mandibles and maxillae are greatly elongate and modified for piercing tissues. The maxillae include a food canal and salivary canals. Both the maxillae and mandibles are enclosed by a sheath-like labium thatprotects the mouthparts when not feeding. Many species in these groups have enlarged muscles and features in the head to provide a strong pumping mechanism called a **cibarial pump**.

Lepidoptera (the butterflies and moths) (Fig. 2.11) have siphoning mouthparts for the uptake of nectar. In this group the maxillae comprise virtually all of the functioning mouthparts. The galeae of the maxillae form a long feeding tube, or proboscis, with a central food canal. The Lepidoptera provide a good example of how morphological features can differ between adults and immatures. Although most adults have siphoning mouthparts, larvae (immatures) have chewing mouthparts.

The feeding apparatus differs among the three major divisions (suborders) of the Diptera. One suborder, Nematocera (Fig. 2.12), has piercing-sucking mouthparts. These consist of a labial sheath surrounding a bundle of stylet-like mouthparts including the labrum-epipharynx, hypopharynx, mandibles, and maxillae. The suborder Brachycera and some species in the suborder Cyclorrhapha (Fig. 2.13) have cutting-sponging mouthparts. The labrum-epipharynx and hypopharynx are modified into cutting organs and together form a food canal. The hypopharynx also includes a salivary canal (Fig. 2.13B). Some species also have blade-like mandibles, although in others the mandibles are lost. The labium provides a groove to contain the labrum-epipharynx and hypopharynx. The distal end of the labium has two lobes called **labella** (singular **labellum**), which are derived from the labial palpi. The labella have many tiny grooves, called **pseudotracheae**, which will draw up blood and other fluids by capillary action. The cutting mouthparts are used to make a wound, and then the labella and the tip of the food canal is inserted into the wound. The labella sponge up blood and draw it to the food canal where it is sucked up. This procedure explains why horse fly and deer fly bites are so painful. Many species in the third suborder, the Cyclorrhapha (Fig. 2.14), have some of the same features of the brachyceran type, but with a much greater development of the labella. In this suborder, feeding occurs though the sponging function of the labella. Consequently, mouthparts of this type are well adapted to semi-solid foods. Additionally, salivary secretions are used to make many solid foods sufficiently liquid for ingestion.

The chewing-lapping mouthparts found in some members of the order Hymenoptera (Fig. 2.15) represent a combination of the chewing and sucking types. A relatively unmodified labrum and mandibles are present and are used to manipulate materials. The maxillae and labium are modified substantially, however. Specifically, the glossa, galea, and labial palpi combine to form a long proboscis for drawing up nectar (Fig. 2.15B).

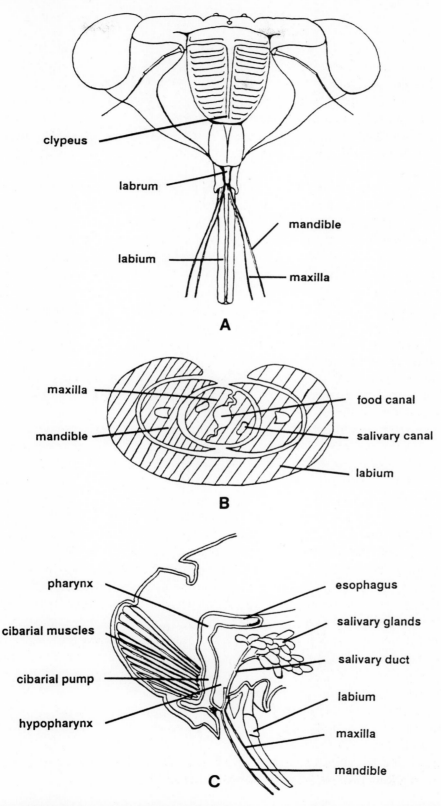

Figure 2.10   Homopteran mouthparts (*Cicada septemdecim*). A. Anterior view. B. Cross section of stylet. C. Lateral interior view. (Modified by KBJ after Matheson 1948)

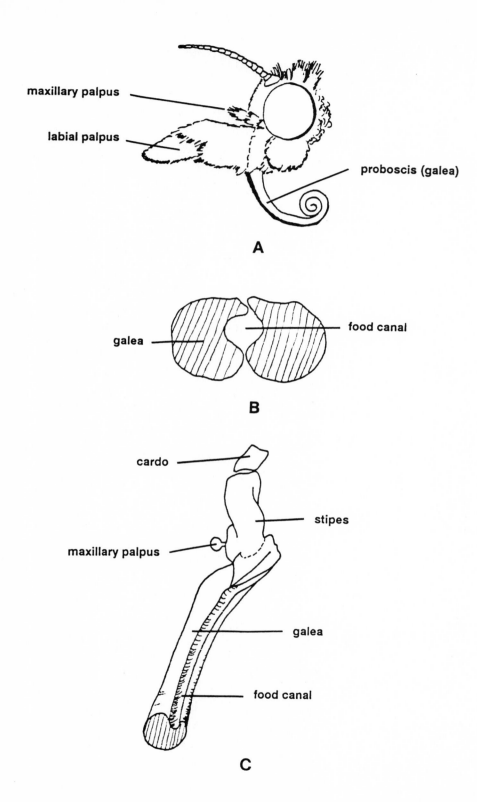

Figure 2.11   Lepidopteran mouthparts (Family Noctuidae). A. Lateral view. B. Cross section of proboscis. C. Detail of proboscis. (Modified by KBJ after Matheson 1948)

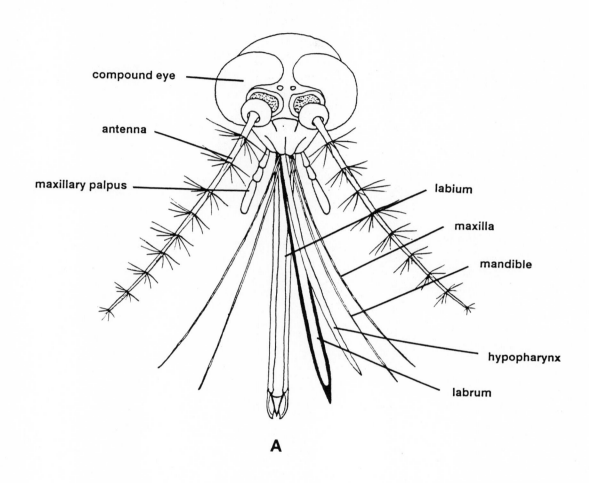

compound eye

antenna

maxillary palpus

labium

maxilla

mandible

hypopharynx

labrum

A

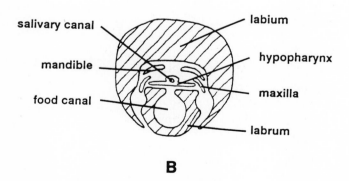

salivary canal

labium

mandible

hypopharynx

food canal

maxilla

labrum

B

Figure 2.12   Dipteran mouthparts (Suborder Nematocera). A. Anterior view. B. Cross section of stylet. (Modified by KBJ after Matheson 1948)

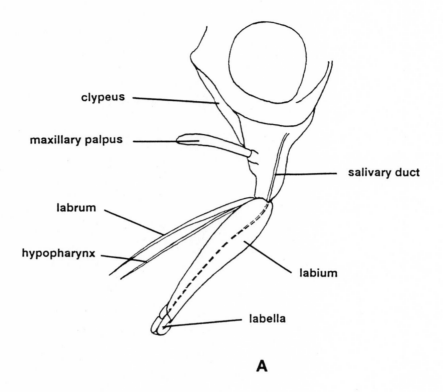

A

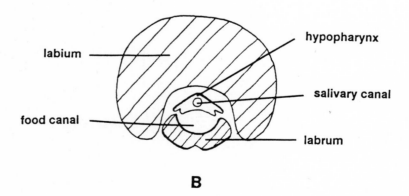

B

Figure 2.13   Dipteran mouthparts (Suborder Brachycera and some Suborder Cyclorrhapha; stable fly, *Stomoxys calcitrans*).   A. Lateral view.   B. Cross section of mouthparts.   (Modified by KBJ after Matheson 1948)

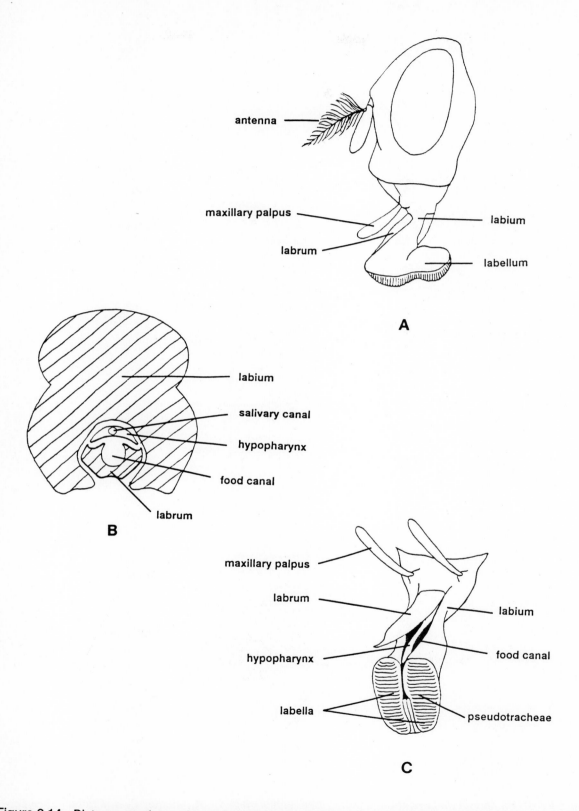

Figure 2.14   Dipteran mouthparts (some Suborder Cyclorrhapha; house fly, *Musca domestica*). A. Lateral view.
B. Cross section of mouthparts. C. Dorso-anterio-lateral view. (Modified by KBJ after Snodgrass 1935)

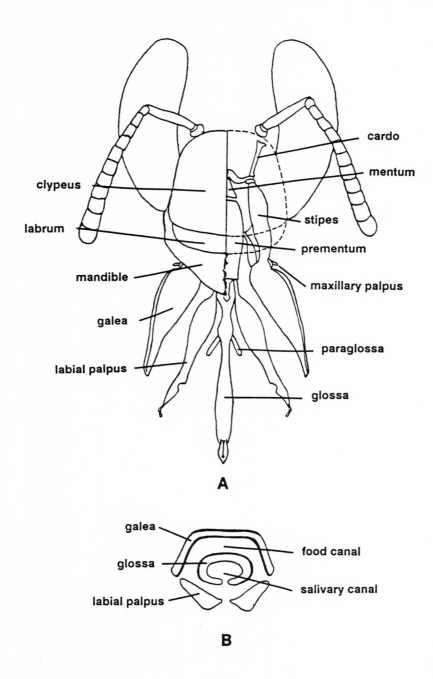

Figure 2.15 Hymenopteran mouthparts (honey bee, *Apis mellifera*). A. Anterior view, with half of clypeus, labrum and left mandible removed to reveal underlying structures. B. Cross section of mouthparts. (Modified by KBJ: A after Matheson 1948; B after Sndograss 1946)

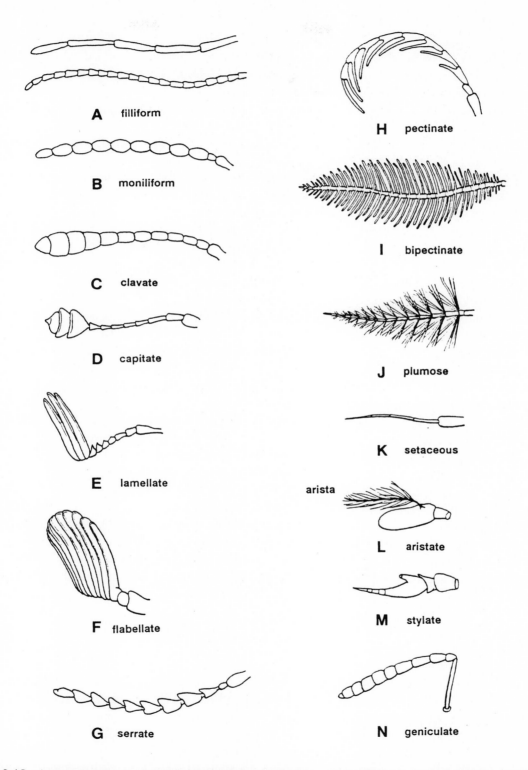

Figure 2.16   Antennal types. A. Filliform (thread-like). B. Moniliform (bead-like). C. Clavate (club-like). D. Capitate (knobbed). E. Lamellate (sheet or leaf-shaped). F. Flabellate (fan-shaped) G. Serrate (saw-toothed) H. Pectinate (single comb-like). I. Bipectinate (double comb-like). J. Plumose (feather-like). K. Setaceous (bristle-like). L. Aristate (with an arista). M. Stylate (with a style or spine). N. Geniculate (elbowed). (KBJ)

Antennae

Figure 2.16 summarizes various antennal types. You need to be familiar with the names of the various types of antennae because these descriptions frequently are used in keys for insect identification. Specific antennal types are:

Filliform (Fig. 2.16A) - thread-like; e.g., Orthoptera, Hemiptera, and some Lepidoptera

Moniliform (Fig. 2.16B) - bead-like, similar to filliform but segments are more rounded or oval; e.g., some Coleoptera

Clavate (Fig. 2.16C) - club-like, increasing in size distally; e.g., some Coleoptera

Capitate (Fig. 2.16D) - knobbed, similar to clavate, but the distal increase in size is more abrupt; e.g., some Coleoptera and some Lepidoptera

Lamellate (Fig. 2.16E) - sheet or leaf-shaped, similar to capitate but the terminal segments are long and flattened; e.g., some Coleoptera

Flabellate (Fig. 2.,16F) - fan-shaped, a type of lamellate with a greater number of elongate, flattened segments; e.g., some Coleoptera

Serrate (Fig. 2.16G) - saw-toothed; some Coleoptera

Pectinate (Fig. 2.16H) - comb-like, teeth arise from one side of segment; e.g., some Coleoptera

Bipectinate (Fig. 2.16I) - double comb-like, teeth arise from two sides of a segment; e.g. some Lepidoptera

Plumose (Fig. 2.16J) - feather-like; e.g., some Diptera and some Lepidoptera

Setaceous (Fig 2.16K) - bristle-like; e.g., Odonata

Aristate (Fig. 2.16L) - having a terminal lobe with anterior bristle (the **arista**); e.g., some Diptera

Stylate (Fig. 2.16M) - having a terminal style or spine; e.g., some Diptera

Geniculate (Fig. 2.16N) - elbowed; e.g., many Hymenoptera

The illustrated types represent standard forms of antennae, however an array of intermediate forms between these major types is possible. Distinctions between some types of antennae can be a matter of interpretation. For example, the filliform and moniliform types are similar and distinguished by the degree of rounding in the flagellar segments. Differentiating the capitate and clavate types depends on how abruptly the distal ends of the antennae are enlarged.

Thorax

Although the thorax always retains its three basic segments, thoracic modifications are common (Fig. 2.17). For example, in the house fly (Fig. 2. 17A), as well as many other Diptera, the thorax is almost entirely mesothorax. Notice how both the pro- and metathorax are reduced and superseded by the mesothorax. The Diptera have only a single pair of wings, and in the suborders Brachycera and Cyclorrhapha the wings and associated musculature are highly developed to allow rapid, strong flight. Consequently, the mesothorax is enlarged to allow for these large flight muscles.

Another example of thoracic alteration is in the Hymenoptera, suborder Apocrita (ants, bees, and wasps) (Fig. 2.17B). The prothorax is reduced and the mesothorax enlarged,

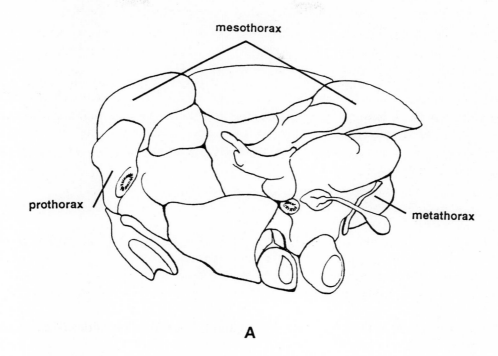

mesothorax

prothorax

metathorax

A

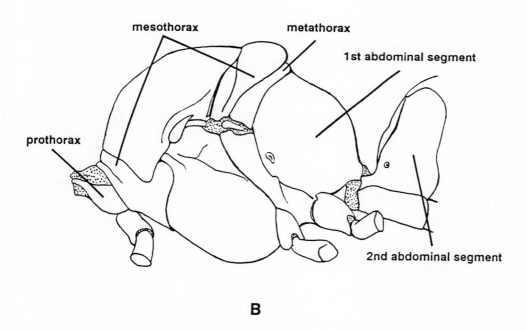

mesothorax

metathorax

1st abdominal segment

prothorax

2nd abdominal segment

B

Figure 2.17 Thoracic modifications. A. Modified thorax in the Diptera (house fly, *Musca domestica*) with enlarged mesothorax. B. Modified thorax in the Hymenoptera (honey bee, *Apis mellifera*) with enlarged mesothorax and first abdominal segment structurally part of the abdomen. (Modified by KBJ: A after West 1951; B after Snodgrass 1946)

however, the most striking modification is that the first abdominal segment is essentially part of the thorax. Other insect groups show similar developments with the first or even first and second abdominal segments becoming more closely associated with the thorax.

Legs

The insect leg can be variously modified. Some modifications occur only to parts of the leg. For example, in the honey bee (Fig. 2.18A) the anterior surface of the hind (metathoracic) tibia is concave and fringed with long hairs to form a pollen basket or pollen storage organ. Scarab beetles, as well as other Coleoptera, have greatly enlarged hind coxae which provide greater strengthening of the thorax (Fig. 2.18B).

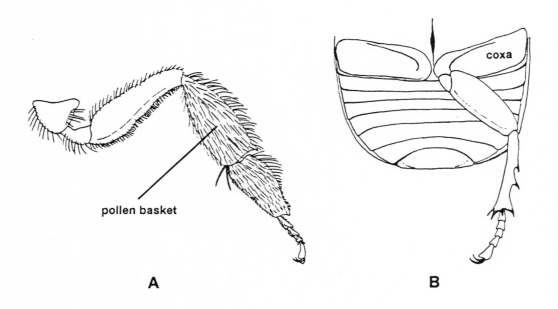

Figure 2.18   Leg modifications. A. Honey bee, *Apis mellifera* metathoracic leg with tibia modified into pollen basket, a pollen storage structure. B. Scarab beetle (Coleoptera, Scarabeidae) with modified coxa of metathoraic which provides structural support for the ventum of the thorax (Modified by KBJ: A after Snodgrass 1946; B after Matheson 1948)

Many leg modifications are associated with a particular function, and identifying the leg type provides an immediate indication of an insect's life history. Figure 2.19 illustrates major leg types. These leg types are:
    Cursorial (Fig. 2.19A, B)- for walking or running
    Saltatorial (Fig. 2.19C, D) - for jumping
    Raptorial (Fig. 2.19E, F) - for grasping prey
    Fossorial (Fig. 2.19G, H) - for digging
    Natatorial (Fig. 2.19I, J) - for swimming
    Scansorial (Fig. 2.19K, L) - for clinging to hairs, common with insect parasites of
mammals

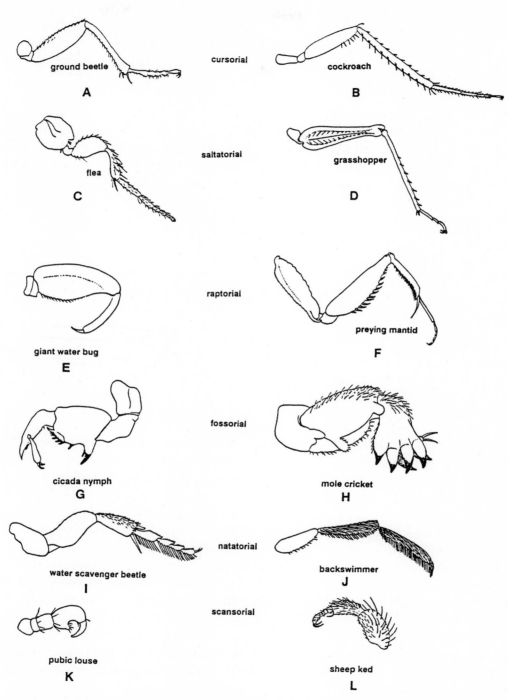

Figure 2.19   Leg types. Cursorial (running or walking): A. Ground beetle (Coleoptera); B. Cockroach (Orthoptera);   Saltatorial (jumping): C. Flea (Siphonaptera); D. Grasshopper (Orhtoptera);   Raptorial (grasping prey): E. Giant water bug (Hemiptera); F. Preying mantid (Orthoptera);   Fossorial (digging): G. Cicada nymph (Homoptera); H. Mole cricket (Orthoptera);   Natatorial (swimming): I. Water scavenger beetle (Coleoptera); J. Backswimmer (Hemiptera);   Scansorial (clinging to hair): K. Pubic louse (Anoplura); L. Sheep ked (Diptera).   (Modified by KBJ: A after Matheson 1948; B,D,E,G after Wilson et al. 1984; D,F,H,I,J after Borror et al. 1976; K after USDHEW-CDC; L after McAlpine et al. 1987)

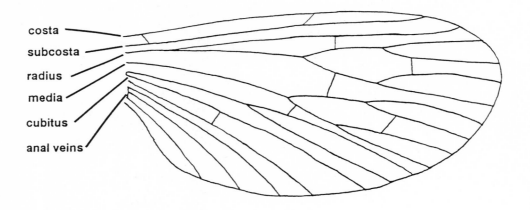

costa
subcosta
radius
media
cubitus
anal veins

Figure 2.20   Characteristic pattern of wing venation and major wing veins of insects. (Modified by KBJ after Borror et al. 1976)

Wings

Just as other other appendages can vary between species, so are the wings subject to alterations. However, the wings of most insect groups are homologous and follow a common pattern of wing venation (Fig. 2.20). Although veins may be lost or vary in shape or location, this pattern or sequential arrangement of veins generally is consistent (Fig. 2.21). Insects wings may have reduced venation and lack some of these veins or some veins may be split, but variations in wing venation generally can be traced to this basic form. However, two insect orders, the Odonata and Ephemeroptera, have venation patterns distinct from other orders (Fig. 2.21A). Differences in wing venation suggests that ancestors of insects in these orders may have evolved wings independently from the common ancestor to other winged insects, or that ancestors of the Odonata and Emphemeroptera diverged from the other orders very early in their evolution.

Besides variation in wing venation, other differences occur (Fig. 2.22). In many groups the fore wing is hardened to function as a protective covering for the hind wing or the hind wing and abdomen. For example, many grasshoppers (Orthoptera) have a hardened, leathery fore wing called a **tegmen** (Fig. 2.22A). This trend is most developed in the Coleoptera, whose fore wings, called **elytra** (singular **elytron**), are completely sclerotized and lack all evidence of venation (Fig. 2.21E). Other modifications include wing hairs (Fig. 2.22D), scales (a characteristic feature of the Lepidoptera) (Fig. 2.22F), and reductions in size. In some Orthoptera a portion of the wing may have small protuberances, called stridulatory pegs, that are used in sound production (Fig. 2.22C). In the Diptera, hind wings are reduced to knobbed structures called **halteres** (singular **halter**). Additionally, species in many orders have non-functional wings or have lost their wings entirely.

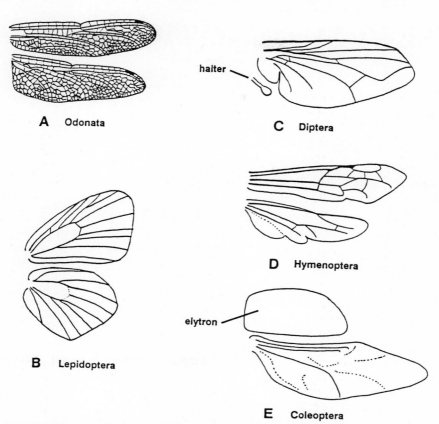

Figure 2.21   Variation in wing venation. A. Odonata. B. Lepidoptera. C. Diptera. D. Hymenoptera. E. Coleoptera. (Modified by KBJ: A,C,D,E after Torre-Bueno 1978; B after Borror et al. 1976)

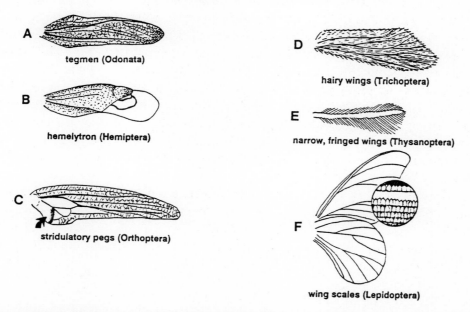

Figure 2.22   Wing modifications. A. Tegmen, a leathery fore wing (Orthoptera); B. Hemelytron, fore wing hardened proximally, with distal membranous region (Hemiptera); C. Stridulatory pegs, structures for sound production (Orthoptera); D. Hairy wings (Trichoptera); E. Narrow, fringed wings (Thysanoptera); F. Scales (Lepidoptera). (KBJ)

3 INSECT SYSTEMATICS

Unquestionably, insects are the most numerous organisms on earth, both in terms of biomass and of species. No exact figures are available for the number of insect species, but Figure 3.1 illustrates one set of estimates that indicates a total of 900,000 insect species or approximately 80% of all animal species (from Swan & Papp 1971). This assessment is almost certainly too conservative. Other authorities have suggested there may be anywhere from 1.5 to 3 million insect species. Investigators studying the insect fauna in the canopies of tropical rain forests have even proposed that as many as 8 or 9 million species of insects may exist. Although this latter estimate may seem extreme, there is no question that substantial numbers of insect species remain to be discovered and described (and, sadly, are threatened with extinction before they are recognized, primarily as a result of habitat destruction).

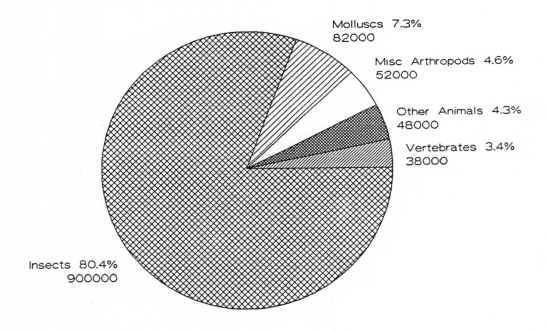

Figure 3.1 Estimated number of animal species by group. (Based on data from Swan and Papp 1971)

Because insects are such a tremendously diverse group of organisms, classifying insects into different orders provides a crucial mechanism for accommodating this diversity. Consequently, insect orders are a primary unit of entomological classification. Recognizing orders and learning characteristics associated with each order is a necessary prerequisite to further identification. However, whereas identification of insects to family, genus, or species may require special expertise, identifying insects to order does not. To study insects for any purpose, including insect pest management, you must be able to identify insects to order.

DISTRIBUTION AND CLASSIFICATION

The distribution of the insects among different orders is, for the most part, as uncertain a question as the total number of insect species. Figures 3.2 and 3.3 present some suggested distributions of the insects worldwide and in North America north of Mexico (based on figures in Borror et al. 1981). Although the proportions of species in the various orders are similar, the numbers of species are not. In fact, by these estimates, North America north of Mexico has only 93,728 described species or 12% of the world's total (762,659). This relative sparseness in North American species can be accounted for by the absence of tropical habitats. Reduced species diversity is characteristic of temperate habitats as compared to tropical environments.

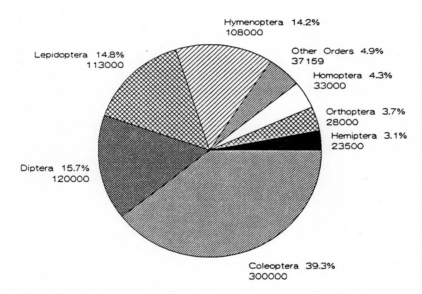

Figure 3.2   Estimated number of described insect species worldwide (Based on data from Borror et al. 1981)

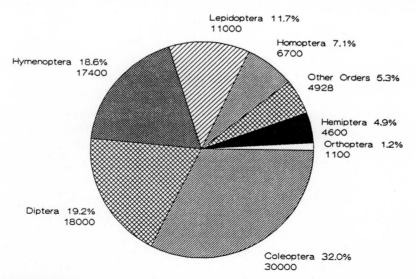

Figure 3.3   Estimated number of described insect species in North America north of Mexico. (Based on data from Borror et al. 1981)

Four orders dominate the Insecta in numbers of species: the Coleoptera, Diptera, Hymenoptera, and Lepidoptera. Not surprisingly, these orders include many insect pests. If we add a further three orders, the Hemiptera, Homoptera, and the Orthoptera, to our list, we can state that most major insect pests are members of one of these seven orders. In terms of economic importance, ecological importance, species diversity, and many other criteria, these are the pre-eminent insect groups. In your studies of insect orders and of individual insect pests, these seven orders merit particular attention.

Although seven orders dominate the Insecta, many other orders exist and must be recognized. These additional orders are of more than academic interest, many include members of economic importance, and some have species of tremendous ecological importance. One of these "minor" orders, the Isoptera or termites, includes species of such importance that they not only provide critical contributions to the breakdown of organic matter in the tropics but also significantly influence gas concentrations in the atmosphere as a result of their activities.

An important reason for the evolutionary success of the insects in general, and the Coleoptera, Diptera, Hymenoptera, and Lepidoptera in particular, is because of differences between adults and immatures. As will be discussed in Chapter 4, the endopterygote insects have immatures and adults that often differ radically in morphology and frequently differ in behavior and ecology as well. As various writers have noted, immature and adult insects may seem as if of different species. This diversity in form and habits enables insects to exploit many habitats and resources that would otherwise be unavailable. These morphological and ecological differences can make the identification of immatures challenging, even identification to order. Nevertheless, correct recognition of immature insects is crucial both because immatures are so frequently encountered and because a great many insect pests are injurious as immatures.

As you might expect, the classification of insects into orders is a matter of interpretation and opinion. We are following the scheme used by Borror et al. (1981) which recognizes 28 orders of insects. These orders are presented in Table 3.1 and are listed to indicate **phylogenetic relationships** (evolutionary arrangements or sequences; genealogies). In different systems, some of the orders we recognize may be combined (e.g., Strepsiptera with Coleoptera) or split into additional orders (e.g., the Orthoptera often is split into Orthoptera [grasshoppers, katydids, and crickets], Dictyoptera [cockroaches and mantids] or Blattodea [cockroaches] and Mantodea [mantids], Grylloblattodea [grylloblattids], and Phasmatodea [walking sticks]). In some schemes certain of the apterygote orders, the Protura, Diplura, and Collembola, are excluded from the insects and considered to be in separate classes.

This distinction between the **Apterygota** (the primitively wingless insects) and the **Pterygota** (the winged or secondarily-wingless insects) represents a major division among orders. Within the pterygote orders, we distinguish between the **Exopterygota** (orders with gradual or incomplete metamorphosis) and the **Endopterygota** (orders with complete metamorphosis). Alternative or additional divisions or distinctions can be drawn within these subgroups. For instance, the Ephemeroptera and Odonata form the **Paleoptera** (old wings). This group lacks a mechanism for flexing the wing over the abdomen, unlike the other pterygote orders, the **Neoptera** (new wings). Additionally, wings in the Paleoptera have numerous cross veins and a venation pattern unlike other orders, which have a common, homologous pattern of wing venation. Presumably, ancestors to the Ephemeroptera and Odonata evolved wings separately from the common ancestor of the Neoptera. Recognizing relationships between orders will assist you in learning about and distinguishing between different insect orders.

Table 3.1 Phylogenetic relationships among insect orders.

Subclass	Division			Order
Apterygota				Protura Collembola Diplura Thysanura Microcoryphia
Pterygota	Exopterygota	Paleoptera		Ephemeroptera Odonata
		Neoptera	Orthopteroid Orders	Orthoptera Isoptera Dermaptera Embioptera
			Hemipteroid Orders	Zoraptera Psocoptera Mallophaga Anoplura Thysanoptera Hemiptera Homoptera
	Endopterygota	Neoptera	Neuropteroid Orders	Neuroptera Coleoptera Strepsiptera
			Panorpoid Orders	Mecoptera Trichoptera Lepidoptera Diptera Siphonaptera
			Hymenopteroid	Hymenoptera

Learning Insect Orders

Reference information at the end of this chapter includes a key to orders for adult and immature insects (Appendix 3.1) and summarizes characteristics of each order (Appendix 3.2). A different key to insect orders is provided in the Pedigo text and the bibliography describes other publications with additional keys. Using more than one key can be helpful in choosing the best characteristics to separate groups, particularly when a couplet uses characters that seem to be obscure or difficult to distinguish. Ultimately, you will be able to identify insects to order by sight, but working specimens through keys is a useful technique for learning distinguishing characters. Additionally, some unusual species may require use of a key to support a sight identification.

Besides the materials included in the manual, you should be familiar with two other references. The standard guide for basic identification of North American insects is "An Introduction to the Study of Insects" by D. J. Borror, D. M. DeLong, and C. A. Triplehorn (entomologists usually refer to it by the name of its authors). Borror, DeLong and Triplehorn includes keys to and descriptions of orders and families for all insects in North America north of Mexico. It is most useful in identifying insects to family, which may not be necessary in identifying most insect pests. However, you should be familiar with Borror, DeLong, and Triplehorn, particularly for those instances when you find an insect that you cannot identify, but you want some idea of what it is. Although you can't get a species identification from Borror, DeLong, and Triplehorn, you can readily determine insects to family. Another useful reference is "A Field Guide to Insects of North America" by D. J. Borror and R. E. White, which is in the Peterson Field Guide series. This field guide includes descriptions of families, but for the most part does not include keys. It is most useful if you have some entomological training, and although not as comprehensive as Borror, DeLong, and Triplehorn, it is an excellent resource for the field. Become acquainted with these references so you can use them when the need arises.

Although there are a large number of insect orders, some can be distinguished by a single character, for example, the enlarged fore tarsi in the Embioptera. Others require a combination of characters, or may include species that lack a primary trait. The Diptera are easily distinguished from all other insects by having a single pair of forewings and short, knobbed structures called halteres in place of hind wings. However, other orders have species with the fore and hind wings connected, so they resemble a single wing. Some orders only have a single pair of wings, although they all lack the halteres as in the Diptera. And within the Diptera itself, a number of species, including at least one insect pest (the sheep ked), are secondarily wingless. Be wary of pitfalls such as these in your identifications. The greatest difficulty is likely to be encountered with the identification of immatures. However, only the larval and pupal insects present substantial difficulty, and the four large orders (Coleoptera, Diptera, Hymenoptera, and Lepidoptera) will account for most of these. Looking at many specimens, working through keys, and recognizing common features and traits are the best approaches for learning the insect orders.

OBJECTIVES

1) Be able to identify adult and immature insects to order.

2) Know the descriptive features, economic importance, and general information on ecology and life history for each of the insect orders.

3) Become familiar with entomological references, in particular, Borror, DeLong, and Triplehorn.

4) Additional objectives (at instructor's discretion):

OPTIONAL DISPLAYS AND EXERCISES

1) Identify insects to order from the array of insects and other invertebrates presented.

2) Try to key out given insect specimens to family.

3) Construct a table of insect orders indicating similarities and differences between orders based on traits that you select.

4) Additional displays or exercises (at instructor's discretion):

TERMS

Apterygota
Endopterygota
Exopterygota

Neoptera
Paleoptera

phylogenetic
    relationship

Pterygota

DISCUSSION AND STUDY QUESTIONS

1) What are the primary morphological traits for distinguishing the insect orders? The primary ecological traits?

2) Which orders include members that are predaceous? parasitic? phytophagous? aquatic?

3) Suggest some reasons for the large species diversity in the Coleoptera, Diptera, Hymenoptera, and Lepidoptera.

4) Generally, larval insects are soft-bodied and less mobile than nymphs and therefore seemingly are more susceptible to predators or other stresses. Yet the endopterygote insects are considered more successful than the exopterygote insects. Explain this apparent contradiction.

BIBLIOGRAPHY

Pedigo, L. P. 1989. Entomology and pest management. Macmillan Pub. Co., New York, NY
    - Chapter 3. Insect classification.

Most of the texts listed in the bibliography for the Introduction include sections on insect orders. In addition to descriptions, some of these may include keys and other identification aids.

General

Borror, D. J., D. M. DeLong, and C. A. Triplehorn. 1981. Introduction to the study of insects. 5th ed. Saunders College Pub., New York, NY
- mentioned in the introduction to this chapter. The standard reference to insects of North America north of Mexico. It includes keys to all families, although use of the keys is easiest with entomological training, particularly a good knowledge of morphology. This is an essential reference.

Borror, D. J., and R. M. White. 1970. A field guide to insects of North America. Houghton Mifflin Co., Boston, MA
- also mentioned in the text. In our opinion the best field guide to North American insects. Other recent field guides include photographs, which we find less useful than drawings and do not focus on insect families (including instead common species). Borror and White includes line drawings and descriptions to family. It is most useful if you have some entomological background.

Chu, H. F. 1949. How to know the immature insects. W. C. Brown Co., Dubuque, IA
- a useful little book that includes keys to immature insects. Descriptions of groups and keys are not as comprehensive as in other references. We have had difficulty with some of the keys.

Commonwealth Scientific and Industrial Research Organization, Division of Entomology. 1971. The insects of Australia. Melbourne University Press, Melbourne
- a comprehensive, beautiful book that includes a number of introductory chapters on general insect features (such as physiology, morphology, behavior, etc.). Besides these beginning chapters, the rest of the book has chapters on each order occurring in Australia. All of the general information and much of the information on individual orders is applicable to North America. A supplemental volume includes more recent information.

Merritt, R. W., and K. W. Cummins (eds.) 1984. An introduction to the aquatic insects of North America. 2nd ed. Kendall Hunt Pub. Co., Dubuque, IA
- an excellent treatment of all groups with aquatic members. It includes keys to adults and immatures, a discussion of features of these groups, and ecological information.

O'Toole, C. 1987. The encyclopedia of insects. Facts on File Publications, New York, NY
- although written for the general reader, this is an excellent survey of the insect orders with information on other important groups of arthropods. The color illustrations and photographs are especially good.

Peterson, A. 1962. Larvae of insects. Part I and II (in separate volumes). Columbus, OH
- for decades the major reference to larval insects. Includes keys and numerous illustrations. It is now replaced by the newly released "Immature Insects".

Richards, O. W., and R. G. Davies. 1977. Imms' general textbook of entomology. 10th ed. Vol 1 and 2. Chapman and Hall, London
  - an important entomological reference. The second volume includes information on the insect orders. Although focused on Great Britain, much of the information is useful for North America.

Stehr, F. W. (ed.) 1987. Immature insects. Vol 1. Kendall Hunt Pub. Co., Dubuque, IA
  - an important new reference to immatures that will include keys to order and families for all North American groups. This volume treats all groups except the Hemiptera, Homoptera, Neuroptera, Coleoptera, Strepsiptera, Diptera, and Siphonaptera which will be covered in a forthcoming second volume.

Swan, L. A., and C. S. Papp. 1972. The common insects of North America. Harper and Row, New York, NY
  - a fine introduction to the insect orders with many good illustrations. A particular emphasis on common and economically important species makes this book especially appealing. Unfortunately out of print.

Specific Groups

Covell, C. V. Jr. 1984. A field guide to moths of eastern North America. Houghton Mifflin Co., Boston, MA
  - one of the Peterson field guide series. An extremely useful guide to moths with good species descriptions and many color plates. Our only complaint is that some plates are in black and white, which makes identification very difficult.

Dillon, E. S., and L. S. Dillon. 1972. A manual of common beetles. Vol 1 and 2. Dover Pub. Inc., New York, NY
  - a useful book with information on many families, genera, and species.

Holland, W. J. 1968. The moth book. Dover Pub. Inc., New York, NY
  - a Dover reprint of Holland's moth book written in the early 1900's. It remains a useful reference to many moth species, and the descriptions including literary quotations are particularly entertaining. Although not available in a Dover edition, Holland wrote a companion volume called "The Butterfly Book".

Howe, W. H. 1975. The butterflies of North America. Doubleday and Co., Garden City, NY
  - although many insect groups lack an adequate reference, there is no shortage of books on butterflies. Other titles emphasize ecology, include photographs, etc.; we like this book because of the clear paintings of all North American species and most subspecies. It also includes good descriptions of families and species with information on habitat, host plants, etc.

McAlpine, J. F. (ed.) 1981 and 1987. Manual of nearctic Diptera. Vol 1 and 2. Agriculture Canada, Research Branch, Ottawa, Ont
  - a superb, essential treatment of North American Diptera. The keys to adults and larvae in Volume 1 are among the finest we've ever used. Chapters on individual fami-

lies include adult (and frequently immature) keys to genera, as well as considerable information on ecology and life history. Extremely well illustrated with virtually all drawings by a single artist. The beautiful illustrations, comprehensive coverage, and the superior keys make these volumes a model against which other references should be measured. A third volume on evolution is under development.

Oldroyd, H. 1966. The natural history of flies. W. W. Norton and Co, Inc., New York, NY
- entomology has been served by some fine writers and this book is a sterling example of such writing. A thoughtful, imaginative discussion of the evolution and ecology of the Diptera.

White, R. E. 1983. A field guide to the beetles. Houghton Mifflin Co., Boston, MA
- another fine volume in the Peterson field guide series. Well illustrated with excellent descriptions of families.

Wilson, E. O. 1971. The insect societies. Belknap Press. Cambridge, MA
- a classic work on the social insects (termites, ants, and bees). Includes information on evolution, behavior, life history and ecology. One of the finest treatments of any group of insects.

Appendix 3.1 - Key for Insect Orders (all stages except eggs)

Format:

couplet #) [prior couplet #]   first description ....couplet # or diagnosis

second description.................................next couplet # or diagnosis

To use:

If stage is unknown use couplet #1, otherwise for adults use #1, for larvae use #58, for naiads use #49, for nymphs use #27, and for pupae use #78.

Adults, Nymphs, and Naiads: couplets 1 - 52

1a)		winged..2
1b)		wingless (or with wings inside membrane over body)....................................27
2a)	[1]	fore wings hardened, heavily thickened, or leathery over entire wing or wing base..3
2b)		fore wings membranous, with scales, or modified into small, knob-like structures..7
3a)	[2]	sucking mouthparts 4
3b)		chewing mouthparts 5
4a)	[3]	fore wing hardened or leathery at base; mouthparts (beak) arising from ventral anterior of head ....**Hemiptera**
4b)		fore wing with uniform texture; mouthparts (beak) arising from ventral posterior of head..**Homoptera**
5a)	[3]	with pincer-like cerci..**Dermaptera**
5b)		without cerci or cerci not modified into pincers.. 6
6a)	[5]	fore wings greatly hardened to form elytra, without venation (although elytra may have lines or punctures) ....**Coleoptera**
6b)		fore wings with branched veins..**Orthoptera**
7a)	[2]	with 2 wings 8
7b)		with 4 wings..13
8a)	[7]	pronotum extends over abdomen..**Orthoptera**
8b)		pronotum does not extend over abdomen 9
9a)	[8]	with thread- or style-like caudal (at end of abdomen) appendages; mouthparts absent or greatly reduced 10
9b)		without caudal appendages; usually mouthparts well-developed.........................11
10a)	[9]	minute (less than 5mm); long antennae; with a long, spine-like caudal appendage (rarely 2 styli); without mouthparts ....**Homoptera**
10b)		moderate to large (greater than 5mm); short, setaceous (bristle-like) antennae; 2 or 3 thread-like caudal filaments ....**Ephemeroptera**

11a) [9]   without any wings modified into knobbed structures; with mandibulate mouth-parts and bulging or swollen face ....**Psocoptera**

11b)   with fore or hind wing modified into knobbed structures (halteres or halter-like organs); face not modified 12

12a) [11]   with well-developed fore wings, hind wings modified into knob-like structures (halteres) ....**Diptera**

12b)   with well-developed hind wings, fore wings reduced to knobbed, halter-like structures ....**Strepsiptera**

13a) [7]   wings very narrow and veinless with extremely long fringe of hairs; mouthparts in conical beak ....**Thysanoptera**

13b)   wings not greatly narrowed and without long fringe of hairs...............................14

14a) [13]   wings completely or partially covered with scales or hairs..................................15
14b)   wings without scales or substantial hairs 16

15a) [14]   mouthparts form coiled proboscis (uncommonly chewing); wings completely or partially covered with scales (uncommonly without scales); body cov-ered with scales or hairs...**Lepidoptera**

15b)   chewing mouthparts with reduced mandibles; wings hairy or with a few scales (not as dense as in Lepidoptera) ....**Trichoptera**

16a) [14]   hind wings smaller (ca. 2/3 or less in length or area) than fore wings.................17
16b)   hind wings approximately equal in size to fore wings 21

17a) [16]   with 2 or 3 caudal filaments..**Ephemeroptera**
17b)   without long abdominal appendages..18

18a) [17]   tarsi with 5 segments; prothorax fused with mesothorax; base of abdomen con-stricted in many groups ....**Hymenoptera**

18b)   tarsi with 2 or 3 segments..19

19a) [18]   with sucking mouthparts; various sizes ....**Homoptera**
19b)   with chewing mouthparts; minute insects 20

20a) [19]   with cerci; face bulging or swollen...**Psocoptera**
20b)   without cerci; face not bulging or swollen...**Zoraptera**

21a) [16]   head modified into very long face or beak...**Mecoptera**
21b)   head not modified into long beak...22

22a) [21]   with short, setaceous (bristle-like) antennae; large compound eyes; wings with numerous cross veins...**Odonata**

22b)   antennae long or short but not setaceous; eyes medium to small in size 23

23a) [22]   hind wings much larger than fore wings; with long cerci..........................**Plecoptera**
23b)   hind wings not greatly larger than fore wings; if cerci present, then short..........24

24a) [23]   medium to large insects; wings with numerous cross veins..................................25
24b)   small insects (less than 8mm); wings without numerous cross veins...................26

25a) [24]   hind legs modified for jumping (hind femora enlarged)......................**Orthoptera**
25b)   hind legs not modified for jumping...**Neuroptera**

26a) [24]   basal segment of fore tarsi greatly enlarged.......................................**Embioptera**
26b)   basal segment of fore tarsi not modified...**Isoptera**

27a) [1]   appendages not covered by membrane...28
27b)   legs and wings under membrane covering, possibly completely covered and
            invisible; usually with compound eyes; a quiescent stage..........(pupae)78

28a) [27]   head modified into very long face or beak....................................**Mecoptera**
28b)   head not modified into long beak...29

29a) [28]   without legs...30
29b)   with legs...32

30a) [29]   mouthparts modified into small needle-like structures; sessile; covered by a
            scale or by waxy secretions..**Homoptera**
30b)   mouthparts not needle-like; not covered by a scale or by wax...........31

31a) [30]   head and thorax fused together; internal insect parasites, sometimes protrude
            between host abdominal segments...**Strepsiptera**
31b)   without compound eyes; generally vermiform (worm-like); head and thorax
            variable: head may be distinct, retracted in thorax, or absent...(larvae)58

32a) [29]   with or without wing pads or ventral abdominal appendages; mouthparts exter-
            nal or retracted into head; 3 thoracic segments can be of different
            forms; tarsi with 1-5 segments; abdominal length variable, some minute
            forms with 12 abdominal segments..33
32b)   without wing pads or ventral abdominal appendages (but may have abdominal
            prolegs); without cerci or abdominal forceps; thoracic legs may be
            reduced to lobes; tarsi usually one segmented; 3 thoracic segments all of
            same form; abdomen usually much longer than head and thorax, never
            with more than 11 abdominal segments ....(larvae)58

33a) [32]   abdomen and thorax narrowly joined...................................**Hymenoptera**
33b)   abdomen and thorax broadly joined...34

34a) [33]   with 3 caudal filaments; usually body with some scales.............................35
34a)   without 3 caudal filaments; with or without scales 36

35a) [34] having cylindrical body with arched thorax; compound eyes are large; with coxal styli ....**Microcoryphia**

35b) body slightly flattened; small compound eyes; without coxal styli ....**Thysanura**

36a) [34] body covered with scales ....**Lepidoptera**
36b) body without scales..37

37a) [36] basal segment of fore tarsi greatly enlarged ....**Embioptera**
37b) basal segment of fore tarsi not modified..38

38a) [37] dorsoventrally or laterally compressed (may be swollen if engorged with blood); ectoparasites of birds and mammals 39

38b) body rounded or not distinctly compressed 43

39a) [38] body laterally compressed ....**Siphonaptera**
39b) body dorsoventrally compressed..40

40a) [39] sucking mouthparts externally visible 41
40b) chewing or sucking mouthparts, not readily visible..42

41a) [40] antennae longer than head ....**Hemiptera**
41b) antennae shorter than head or absent ....**Diptera**

42a) [41] head broader than thorax at base; chewing mouthparts ....**Mallophaga**
42b) head slender and pointed, narrower than thorax; sucking mouthparts...**Anoplura**

43a) [38] without antennae; 12 abdominal segments ....**Protura**
43b) with antennae; fewer than 12 abdominal segments..44

44a) [43] with collophore (ventral, eversible tube on first abdominal segment); usually tenaculum (catch) and furcula (springing organ) on third and fourth abdominal segments,respectively ....**Collembola**

44b) without collophore, tenaculum, or furcula 45

45a) [44] with 2 caudal (at end of abdomen) appendages (appendages longer than cerci); minute..**Diplura**

45b) without 2 caudal appendages..46

46a) [45] with sucking mouthparts..47
46b) with chewing mouthparts 48

47a) [46] either without distinct head or with head having mouthparts arising from ventral posterior of head..**Homoptera**

47b) with mouthparts arising from ventral anterior of head............................**Hemiptera**

48a) [46] aquatic; frequently with tracheal gills ....(naiads)49
48b) terrestrial; without gills..51

49a) [48] with hinged labium, modified for grasping prey; large..............................**Odonata**
49b)   without modified labium 50

50a) [49] with lateral, leaf-like gills on the abdomen; 3 caudal (at end of abdomen) fila-
            ments...**Ephemeroptera**
50b)   usually with filamentous thoracic gills near leg bases; 2 caudal (at end of
            abdomen) filaments...**Plecoptera**

51a) [48] with abdominal forceps or claspers 52
51b)   without forceps or claspers 53

52a) [51] body dark and heavily sclerotized (hardened)..............................**Dermaptera**
52b)   body soft and white, except for hardened forceps ....**Diplura**

53a) [51] sclerotized (hardened) body, frequently dark...54
53b)   soft body, usually white or pale..55

54a) [53] small insects (less than 5 mm); narrow body; mouthparts in conical beak;
            without tarsal claws...**Thysanoptera**
54b)   medium to large sized insects (greater than 5 mm); with tarsal claws; relatively
            unmodified chewing mouthparts ....**Orthoptera**

55a) [53] with cerci..56
55b)   without cerci...57

56a) [55] with 2 tarsal segments...**Zoraptera**
56b)   with 4 tarsal segments...**Isoptera**

57a) [55] antennae long (greater than one third of body length); face appears bulging or
            swollen...**Psocoptera**
57b)   antennae short (less than one fourth of body length); face not bulging or swol-
            len...**Strepsiptera**

Larvae: couplets 58 - 77

58a) [31   without thoracic legs 59
58b) &32] with thoracic legs or legs reduced to lobes...66

59a) [58] without distinct head or head partially sclerotized (hardened) or partially
            concealed within thorax..60
59b)   with distinct, well-sclerotized head...62

60a) [59] normal chewing mouthparts, with opposable mandibles and maxillae; usually
            with antennae..**Coleoptera**
60b)   mouthparts reduced or substantially modified, only mandibles (not maxillae)
            opposable or mouthparts not opposable (move parallel); usually without
            antennae 61

| 61a) | [60] | mouthparts may be reduced to a pair of hooks or absent; various body forms often vermiform (worm-like); sometimes with fleshy protuberances, gills, or posterior abdominal segment elongated and modified into breathing tube..**Diptera** |
| 61b) | | mandibulate mouthparts greatly or slightly reduced to sharp pointed mandibles, hardened plates, or a fleshy opening; larvae usually Scarabaeiform (u-shaped)..**Hymenoptera** |

| 62a) | [59] | face with adfrontal area (triangular region between eyes defined by sutures); labium with spinneret (finger-like silk producing organ); usually with abdominal prolegs bearing crochets (tiny, curved hooks) ......**Lepidoptera** |
| 62b) | | without adfrontal area, crochets, or spinneret ....63 |

| 63a) | [62] | with long, dorsal, abdominal setae that increase in length on posterior segments; 11 abdominal segments ....**Siphonaptera** |
| 63b) | | without setae of increasing length; 9 or 10 abdominal segments.........................64 |

| 64a) | [63] | without spiracles on most abdominal segments; mandibulate mouthparts, mouthparts reduced to hooks, or mandibles modified into mouth brushes; body forms variable; includes aquatic, parasitic, and phytophagous species ....**Diptera** |
| 64b) | | with spiracles on thorax and most abdominal segments; normal mandibular mouthparts ....65 |

| 65a) | [64] | one pair of thoracic spiracles..**Coleoptera** |
| 65b) | | 2 pairs of thoracic spiracles...**Hymenoptera** |

| 66a) | [58] | with thoracic legs reduced to lobes...67 |
| 66b) | | with thoracic legs...69 |

| 67a) | [66] | face with adfrontal area (triangular region between the eyes defined by sutures); usually with abdominal prolegs bearing crochets (tiny, curved hooks) ....**Lepidoptera** |
| 67b) | | without adfrontal area or prolegs with crochets ....68 |

| 68a) | [67] | body straight and of uniform diameter ....**Hymenoptera** |
| 68b) | | body scarabeiform (u-shaped) with mid-abdominal diameter greater than that of either end..**Coleoptera** |

| 69a) | [66] | with abdominal prolegs...70 |
| 69b) | | without prolegs...72 |

| 70a) | [69] | head with 10 or more ocelli; end of abdomen resembles sucking disk.**Mecoptera** |
| 70b) | | head with fewer than 10 ocelli; end of abdomen may include anal proleg but does not resemble sucking disk ....71 |

71a)   [70]   with 6 or more pairs of abdominal prolegs...**Hymenoptera**
71b)   with 5 or fewer pairs of abdominal prolegs...**Lepidoptera**

72a)   [69]   head hypognathous (directed downward)...73
72b)   head prognathous (directed forward)...74

73a)   [72]   fore legs smaller than others; middle and hind legs project laterally....**Mecoptera**
73b)   legs of approximately equal size...**Coleoptera**

74a)   [72]   tarsi with one claw..75
74b)   tarsi with 2 claws...77

75a)   [74]   with long needle-like mouthparts; aquatic insects that feed on freshwater
                sponges...**Neuroptera**
75b)   with normal chewing mouthparts...76

76a)   [75]   thoracic legs elbowed; sometimes with abdominal gills..........................**Coleoptera**
76b)   aquatic; caudal (at end of abdomen) hooks ; gills may be present on thorax and
                abdomen..**Trichoptera**

77a)   [74]   without distinct labrum and clypeus, mandibles and maxillae modified to form
                sickle-shaped sucking apparatus; or with distinct labrum and clypeus
                and chewing mouthparts (possibly with lateral thoracic or abdominal
                gills ....**Neuroptera**
77b)   without distinct labrum and clypeus; normal chewing mouthparts or sickle-
                shaped mandible with internal tube or external groove but unmodified
                maxillae ....**Coleoptera**

Pupae: couplets 78 - 86

78a)   [27]   obtect (appendages invisible or tightly fastened to body wall)............................79
78b)   exarate (appendages not secondarily attached or held to body).........................80

79a)   [78]   all appendages invisible, shape ellipsoid with latitudinal rings (coarctate - pupae
                inside final larval integument, the puparium); or with visible append-
                ages (only 2 wings) and respiratory tubes.......................................**Diptera**
79b)   with visible appendages (4 wings but fore wings may conceal hind wings) but
                without respiratory tubes..**Lepidoptera**

80a)   [78]   laterally compressed...**Siphonaptera**
80b)   not laterally compressed...81

81a)   [80]   functional mandibles (darkened, hardened, and movable)................................82
81b)   mandibles not apparent or not functional (immovable)....................................85

82a)   [81]   pronotum shorter than mesonotum or indistinct..83
82b)   pronotum obvious and as long as mesonotum 84

| 83a) | [82] | abdomen with dorsal hooks; frequently with filamentous thoracic gills; usually within stone or twig case; aquatic..**Trichoptera** |
| 83b) | | abdomen without hooks or gills...**Lepidoptera** |

| 84a) | [82] | hypognathous (mouthparts directed downward); head modified into long face or beak...**Mecoptera** |
| 84b) | | prognathous (mouthparts directed forward); without head modified into beak or long face...**Neuroptera** |

| 85a) | [81] | metanotum much larger than rest of thorax ....**Strepsiptera** |
| 85b) | | metanotum no more than equal in length to rest of thorax...................................86 |

| 86a) | [85] | pronotum shorter than mesonotum...**Hymenoptera** |
| 86b) | | pronotum at least equal in length to mesonotum; possibly with rudimentary elytra..**Coleoptera** |

Appendix 3.2 - Descriptions of Insect Orders

Anoplura (*anopl* = unarmed; *ura* = tail) **sucking lice**

Description and Identification
> Metamorphosis: gradual
> Immature: like adult

> Adult:
>> Mouthparts - sucking
>> Tarsal Segments - 1
>> Size - minute
>> Wings - wingless
>> Other Characteristics - head usually slender and pointed; may lack eyes; scansorial legs (adapted to climb on hairs)
> Similar Orders: Mallophaga (wide head, chewing mouthparts, 1-2 segmented tarsi)

Habitat
> ectoparasites of mammals, with entire life on host

Economic Importance
> ectoparasites of man and livestock; vector some diseases

Classification and Groups
> North America: 9 families, 70 species
> Important Families:
>> Pediculidae - includes human head and body louse
>> Pthiridae - single member in North America is human pubic, or crab, louse

Notes:

A

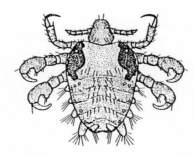

B

Figure 3.4   Anoplura. A. Rodent louse (Haematopinidae). B. Pubic louse (Phthiridae). (Courtesy USDA)

Coleoptera (*coleo* = sheath; *ptera* = wings)   **beetles**

Description and Identification
Metamorphosis: complete

Immature: larvae extremely variable in form (including virtually all larval types although campodeiform, eruciform, and scarabeiform common), all have head usually with chewing mouthparts without obvious labrum, 3 segmented thorax usually with segmented legs and one thoracic spiracle; pupae adecticous with most exarate but some obtect

Adult:

    Mouthparts - chewing

    Tarsal Segments - usually 3 to 5

    Size - minute to large

    Wings - 4 (a few species wingless); fore wings greatly hardened and modified into elytra, hind wings membranous and folded under elytra

    Other Characteristics: antennae usually 11 segments but can be from 0-8, extremely variable in form

Similar Orders:

    Immatures - Hymenoptera (2 thoracic spiracles), Neuroptera (mandibles and maxillae modified for sucking or if chewing mouthparts have an obvious labrum), Trichoptera (have posterior claws);

    Adults - Dermaptera (cerci modified into pincers), Hemiptera and Homoptera (mouthparts modified into beak for sucking)

Habitat
all habitats, most phytophagous but some species insect predators, aquatic, feed on fungi or mammalian parasites

Economic Importance
extremely important, include major plant pests

Classification and Groups
North America: 112 families, 30,000 species

Important Families:

    Carabidae - ground beetles; most predaceous

    Chrysomelidae - leaf beetles; many serious plant pests

    Coccinelidae - ladybird beetles; most predaceous (on aphids), a few plant pests

    Curculionidae - weevils; many important pests both of cultivated plants and stored products

    Dermestidae - dermestid beetles; household and stored product pests

    Elateridae - wireworms (larvae), click beetles (adults); some species plant pests

    Scarabaeidae - scarab beetles (includes June beetles); larvae (white grubs) plant pests

    Scolytidae - bark beetles; many major forest pests

    Staphylinidae - rove beetles; most predaceous and a few parasitic species

    Tenebrionidae - darkling beetles; a number of significant stored grain pests

Notes:

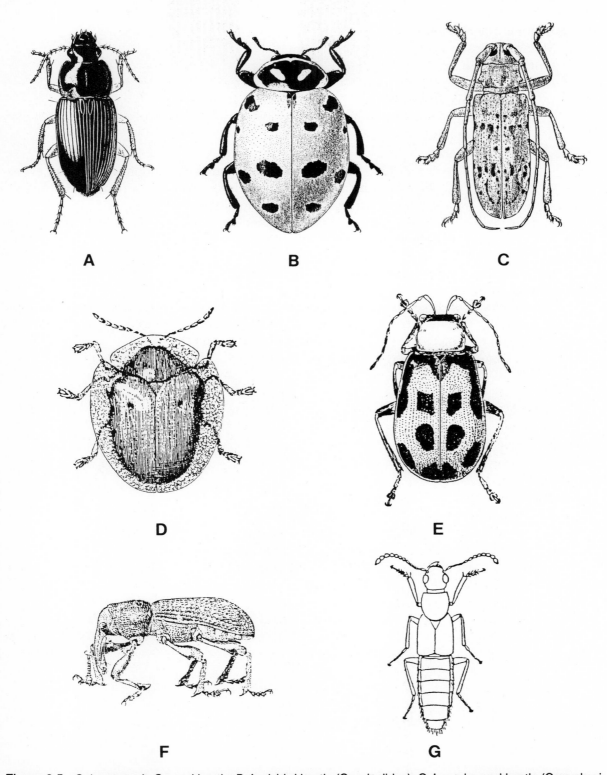

Figure 3.5   Coleoptera. A. Ground beetle. B. Ladybird beetle (Coccinelidae). C. Long-horned beetle (Cerambycidae). D. Tortoise beetle (Chrysomelidae). E. Bean leaf beetle (Chrysomelidae). F. Pales weevil (Curculionidae). G. Rove beetle (Staphylinidae). (Courtesy USDA)

Collembola (*coll* = glue; *embola* = wedge)   **springtails**

Description and Identification
Metamorphosis: none
Immature: like adult
Adult:
Mouthparts - chewing
Tarsal Segments - 1
Size - minute
Wings - none
Other Characteristics - first abdominal segment with collophore (ventral, eversible tube); spring apparatus consisting of tenaculum (catch) and furcula (spring) on third and fourth abdominal segments, respectively;
Similar Orders: Protura, Diplura, wingless species of other orders, but these lack a central collophore
Habitat
organic debris, green plants, and soil (very abundant)
Economic Importance
occasionally pests in gardens or greenhouses, on mushrooms; reported as pest of alfalfa in Australia
Classification and Groups
North America: 5 families, 315 species
Notes:

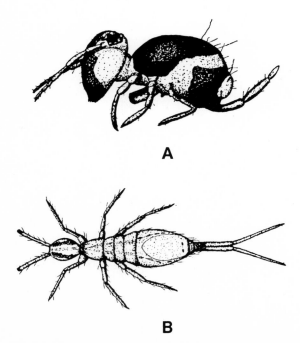

A

B

Figure 3.6   Collembola - springtails. A. Sminthuridae. B. Entomobryidae. (LPP)

Dermaptera   *(derma* = skin; *ptera* = wings)   **earwigs**

Description and Identification
> Metamorphosis: gradual
> Immature: like adult, but wingless
> Adult:
>> Mouthparts - chewing
>> Tarsal Segments - 3
>> Size - small to medium
>> Wings - 4 or none; fore wing short and thickened, hind wings membranous and folded
>> Other Characteristics - cerci modified into pincers
> Similar Orders: Coleoptera (rove beetles [Staphylinidae] resemble Dermaptera, but lack pincer-like cerci)

Habitat
> debris; nocturnal

Economic Importance
> little importance; may injure some cultivated plant and may be minor household pest

Classification and Groups
> North America: 5 families, 20 species

Notes:

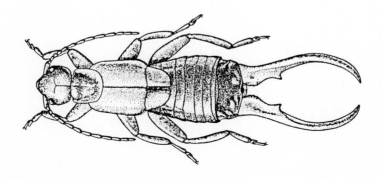

Figure 3.7   Dermaptera - European earwig (Forficulidae). (Courtesy USDA)

Diplura   (*di* = two; *plura* = tails)   **diplurans**

Description and Identification

> Metamorphosis: none
> Immature: like adult
> Adult:
>> Mouthparts - chewing
>> Tarsal Segments - 1
>> Size - minute
>> Wings - none
>> Other Characteristics - lack eyes; 2 appendages at end of abdomen
> Similar Orders - Protura (lacks antennae), Microcoryphia and Thysanura (have 3 abdominal appendages [caudal filaments] and scales)

Habitat

> debris, under bark, and similar damp places

Economic Importance

> none

Classification and Groups

> Order called Entotrophi by some workers
> North America: 4 families, 57 species

Notes:

Diptera   (*di* = two; *ptera* = wings)   flies, gnats, midges, and mosquitoes

Description and Identification
  Metamorphosis: complete
  Immature: variable; larvae are legless eucephalic (recognizable head) in the suborder
    Nematocera, hemicephalic (half of head recognizable) in the Brachycera and
    acephalic (without recognizable head) in the Cyclorrhapha (vermiform larvae),
    in these latter suborders larval mouthparts are reduced to a pair of mouth
    hooks; nematocerous and most brachycerous pupae are adecticous obtect,
    cyclorrhaphous pupae are coarctate (adecticous exarate inside the last larval
    integument, the puparium)
  Adult:
    Mouthparts - sucking (sponging)
    Tarsal Segments - 5
    Size - minute to large
    Wings - 2 (few species with none); membranous fore wings, hind wings modi-
      fied into small knobbed structures (halteres)
    Other Characteristics - antennae variable (Nematocera: filiform or plumose;
      Brachycera: short, horn-like; Cyclorrhapha: aristate); generally, Nema-
      tocera slender, fragile flies, other suborders more robust flies
  Similar Orders: Hymenoptera (but have 4 wings)
Habitat
  virtually all habitats, some in vegetation, aquatic, insect predators or parasites, feed or
  ectoparasites of vertebrates
Economic Importance
  most important order with respect to human health and disease transmission, addition-
  ally some species plant pests and others biological control agents
Classification and Groups
  North America: 107 families, 18,000 species
  Important Subdivisions (Suborders): Nematocera, Brachycera, and Cyclorrhapha
    (some authorities combine the last two suborders or have an otherwise differ-
    ent treatment of higher classification)
  Important Families:
    Agromyzidae - leaf miner flies; some larvae plant pests (mine in leaves)
    Anthomyiidae - anthomyiid flies; larvae of some species important plant pests
    Cecidomyiidae - gall gnats; many larvae phytophagous causing gall formation;
      some plant pests including Hessian fly
    Culicidae - mosquitoes; the most important group of insects with respect to
      human health
    Muscidae - muscid flies; includes household (e.g., house fly) and livestock (face
      fly) pests
    Syrphidae - syrphid flies; many larvae predaceous
    Tabanidae - horse and deer flies
    Tachinidae - tachinid flies; important biological control agents (larvae insect
      parasites)
    Tephritidae - fruit flies; includes important fruit pests such as apple maggot and
      Mediterranean fruit fly
Notes:

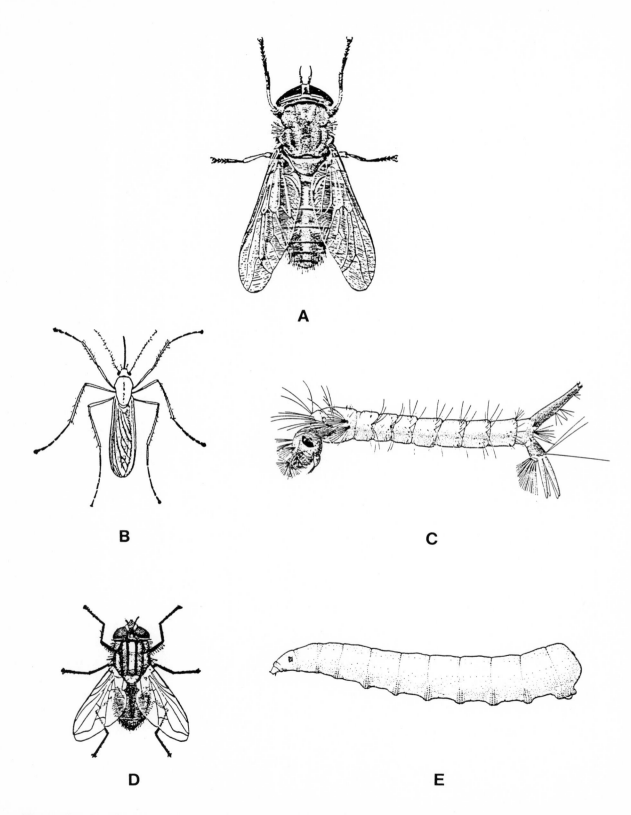

Figure 3.8   Diptera. A. Striped horse fly (Tabanidae). B. Mosquito (Culicidae) C. Mosquito larva (Culicidae). D. House fly (Muscidae). E. House fly larva (Muscidae). (Courtesy USDA)

Embioptera *(embio* = lively; *ptera* = wings)   **webspinners**

Description and Identification

        Metamorphosis: gradual

        Immature: like adult

        Adult:

                Mouthparts - chewing

                Tarsal Segments - 3

                Size - small

                Wings - 4 or none

                Other Characteristics - basal segment of front tarsi greatly enlarged and contains silk producing gland; cerci present;

        Similar Orders: Isoptera, Psocoptera, and Zoraptera (none with enlarged basal tarsi segments)

Habitat

        in soil debris, moss or lichens

Economic Importance

        none

Classification and Groups

        North America: 3 families, 9 species

Notes:

Ephemeroptera   (*ephemera* = short-lived; *ptera* = wings)   **mayflies**

Description and Identification
> Metamorphosis: incomplete
> Immature: naiads; leaf-like gills laterally on abdomen; 3 long caudal filaments (resemble tails); feed on aquatic organisms and organic debris; a winged subimago (sexually immature, preadult stage) emerges from water and usually molts following day
> Adult:
>> Mouthparts - none (vestigial)
>> Tarsal Segments - 3 to 5
>> Size - moderate
>> Wings - usually 4, but some families with 2; wings membranous with net-like cross veins; large triangular fore wings, small hind wings (if present)
> Other Characteristics - small, setaceous antennae; 2 long caudal filaments (tails); adults very short-lived (1-2 days)
> Similar Orders:
>> Adults - Odonata (hind wing as large as fore wing), Plecoptera (long antennae)
>> Immatures: Odonata (without lateral gills on abdomen), Plecoptera (without lateral gills or 3 caudal filaments)

Habitat
> immatures aquatic, adults near water

Economic Importance
> indicators of water quality; mass emergence of adults can be large enough to cause problems (cover roads, block car radiators, etc.)

Classification and Groups
> North America: 17 families, 606 species

Notes:

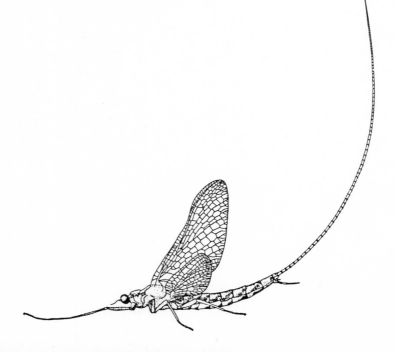

Figure 3.9   Ephemeroptera. Adult Heptageniidae. (Courtesy USDA)

Hemiptera (*hemi* = half; *ptera* = wings)   **true bugs**

Description and Identification

Metamorphosis: gradual

Immature: like adult

Adult:

      Mouthparts - sucking

      Tarsal Segments - 2 or 3

      Size - medium

      Wings - 4 or none; fore wing thickened at base and membranous at tip

      Other Characteristics - mouthparts form beak arising from tip of head; antennae either long (5 segments or less) and obvious or short and concealed

Similar Orders: Coleoptera (fore wing entirely hardened into elytron), Homoptera (fore wing entirely membranous, beak arises from posterior of head)

Habitat

some aquatic species, most on vegetation, a few species ectoparasites of birds and mammals, other species insect predators

Economic Importance

of major importance; many species significant pests of cultivated crops, some species vectors of human disease, some species beneficial as insect predators

Classification and Groups

Some authorities combine Homoptera and Hemiptera, with suborders Heteroptera (which includes Hemiptera as defined here) and Homoptera; in such a scheme the combined order is called Hemiptera

North America: 44 families, 4,600 species

Important Families:

      Cimicidae - bed bugs

      Lygaeidae - seed bugs; large family with many plant pest species (e.g., chinch bug)

      Miridae - plant or leaf bugs; largest family in order; most species phytophagous, but some predaceous; includes important plant pests (e.g.,tarnished plant bug)

      Nabidae - damsel bugs; insect predators

      Pentatomidae - stink bugs; phytophagous and predaceous species; pests include Harlequin bug and green stink bug

      Reduviidae - assassin and conenose or kissing bugs; insect predators, but one group (conenose bugs - subfamily Triatominae feeds on man and vectors an important human pathogen in South America (causing Chagas' disease)

Notes:

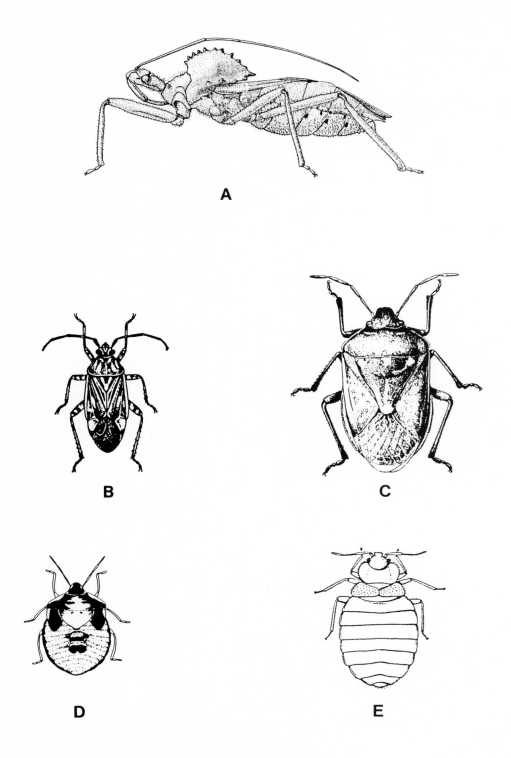

Figure 3.10 Hemiptera. A. Assassin bug (Reduviidae). B. Lygus bug (Miridae). C. Stink bug (Pentatomidae). D. Stink bug nymph (Pentatomidae). E. Bed bug (Cimicidae). (A,B,D,E courtesy USDA; C courtesy Illinois Cooperative Extension Service)

Homoptera (*homo* = alike; *ptera* = wings)

aphids, cicadas, hoppers, psyllids, scales, and whiteflies

Description and Identification

Metamorphosis: gradual, however some groups with more complicated scheme

Immature: generally like adult but without wings

Adult:

Mouthparts - sucking

Tarsal Segments - 1 to 3

Size - minute to large

Wings - 4, 2, or none; when present wings membranous and of uniform texture

Other Characteristics - mouthparts modified to form beak that arises from back of head; antennae may be short and setaceous or long and filiform

Similar Orders: Coleoptera (elytra and chewing mouthparts), Hemiptera (half thickened fore wing and beak arising from front of head)

Habitat

on vegetation, all species phytophagous

Economic Importance

very important, order includes many major plant pests

Classification and Groups

This order is included with the Hemiptera by some authorities (see Hemiptera for details)

North America: 38 families, 33,000 species

Important Families:

Aleyrodidae - whiteflies; important citrus and greenhouse pests

Aphididae - aphids; many significant plant pests, some vector plant diseases

Cicadellidae - leafhoppers; a number of plant pests, plant disease vectors

Cicadidae - cicadas; usually minor injury but oviposition in plant tissues is damaging to young trees

Coccidae - soft scales; citrus and greenhouse pests

Diaspididae - armored scales; some important orchard pests including San Jose and oystershell scales

Pseudococcidae - mealybugs; important greenhouse pests

Psyllidae - psyllids; important greenhouse pests

Notes:

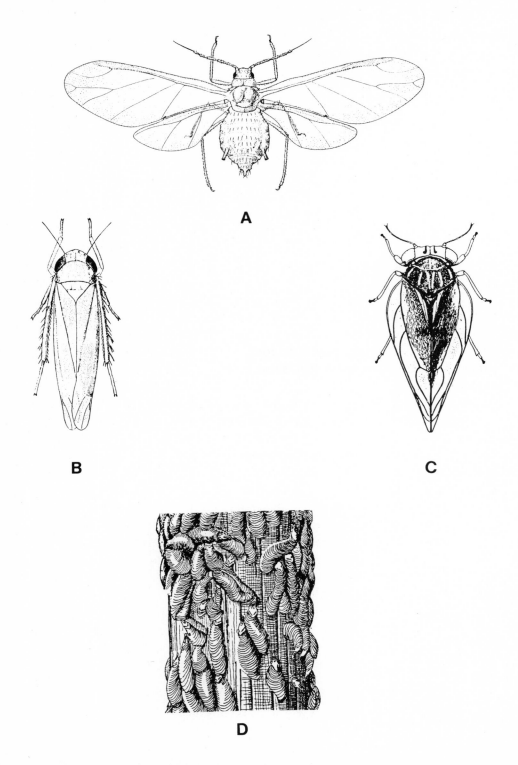

Figure 3.11   Homoptera. A. Aphid (Aphididae). B. Potato leafhopper (Cicadellidae). C. Pear psylla (Psyllidae). D. Oystershell scales on tree (Diaspididae). (A,C,D Courtesy USDA; B Courtesy Illinois Cooperative Extension Service)

Hymenoptera   (*hymen* = a membrane; *ptera* = wings)   **ants, bees, sawflies, and wasps**

Description and Identification
Metamorphosis: complete

Immature: larvae of suborder Apocrita eucephalic but vermiform, of Symphyta eruci- form with 6 prolegs and 1 large pair of ocelli; most pupae are adecticous exa- rate and pupate inside a silken cocoon (some parasitic species adecticous ob- tect)

Adult:

Mouthparts - chewing, chewing/sucking

Tarsal Segments - 5

Size - minute to large

Wings - 4 or none; fore wing larger than hind wing, wing coupling mechanisms to form a single functional wing

Other Characteristics - Symphyta: antennae filiform and usually long, base of abdomen broadly joined to thorax; Apocrita: base of abdomen con- stricted in segment joining thorax (propodeum), larvae of many species insect parasites, antennae variously modified

Similar Orders: Diptera (2 wings and halteres)

Habitat
various; on vegetation, phytophagous species with larvae in or on plants, parasitic species with larvae in or on host, social species in colonies

Economic Importance
some plant pests; honey production; primarily important in biological control of insects and pollination; some stinging Hymenoptera can cause medical problems

Classification and Groups
North America: 74 families, 93,728 species

Important Subdivisions: Apocrita (ants, bees, and wasps) and Symphyta (sawflies)

Important Families:

Apidae - bumble bees and honey bees

Braconidae - braconid wasps

Cephidae - stem sawflies

Formicidae - ants

Ichneumonidae - ichneumonid wasps

Tenthredinidae - tenthredinid sawflies

Vespidae - paper wasps, yellowjackets, and hornets

Notes:

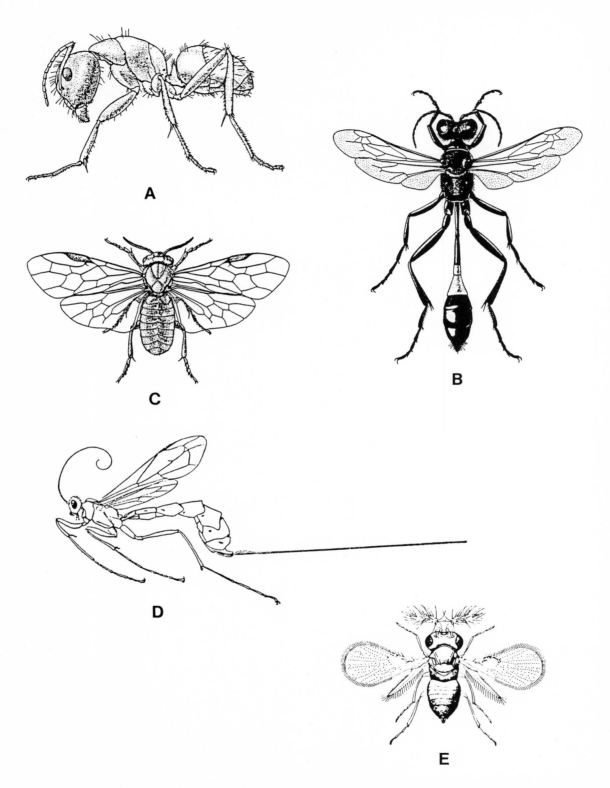

Figure 3.12   Hymenoptera. A. Common ant (Formicidae). B. Thread-waisted wasp (Sphecidae). C. Pine sawfly (Diprionidae). D. *Trichogramma* wasp (Trichogrammatidae). E. Ichneumonid wasp (Ichneumonidae).   (Courtesy USDA)

Isoptera   *(iso* = equal; *ptera* = wings)   **termites**

Description and Identification

Metamorphosis: gradual

Immature: like adult

Adult:

Mouthparts - chewing, but can be greatly modified

Tarsal Segments - 4

Size - small

Wings - reproductives with 4 wings that are broken off after dispersal flight; other castes wingless; wings long and narrow with fore and hind wings similar

Other Characteristics - only head heavily sclerotized; body white with flexible thin exoskeleton; thorax and abdomen broadly joined; antennae relatively short; different castes (as many as 5) with distinctly different morphologies, behaviors, and functions in colony including soldiers, workers, and reproductives

Similar Orders: Embioptera (basal segment of fore tarsi enlarged), Hymenoptera (narrow joint between thorax and abdomen), Psocoptera (tarsi 2-3 segmented, antennae long and slender), Zoraptera (tarsi 2 segmented)

Habitat

in wood or soil; live in colonies or nests

Economic Importance

very important pests; destroy wood (including wooden features of buildings, fences) and wood products (e.g., books); however, also beneficial because extremely important ecologically in contributing to the breakdown of plant debris

Classification and Groups

North America: 4 families, 41 species

Notes:

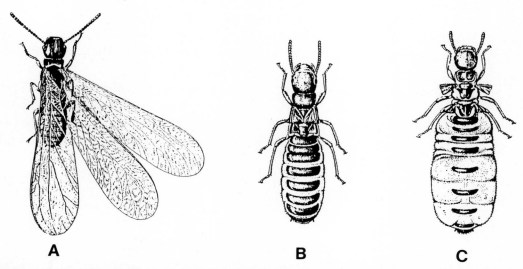

Figure 3.13   Isoptera - eastern subterranean termites (Rhinotermitidae). A. Worker. B. King. C. Queen. (Courtesy USDA)

Lepidoptera   (*lepido* = scale; *ptera* = wings)   **butterflies** and **moths**

Description and Identification
Metamorphosis: complete

Immature: larvae eruciform, well-developed head usually with a short antennae and 6 ocelli on each side, generally 10 abdominal segments frequently with 5 or fewer pairs of prolegs, most larvae are phytophagous; most pupae are adecticous obtect but a few minor groups are decticous (exarate), pupation often occurs inside a silken cocoon

Adult:
Mouthparts - sucking (proboscis)

Tarsal Segments - 5

Size - small to large

Wings - 4 (a few wingless); wings covered with scales and may be variously colored or patterned; hind wings smaller than fore wings, both coupled together (by various mechanisms) in flight to form a single functional wing

Other Characteristics - like wings, body often covered with scales; antennae long and capitate (butterflies), clavate, filiform, or sometimes plumose; usually lack maxillary palps; compound eyes large with obvious facets; most adults feed on nectar or other liquids, some do not feed; generally, butterflies diurnal, moths nocturnal or crepuscular

Similar Orders: Trichoptera (lack wing scales)

Habitat
on vegetation

Economic Importance
order includes many severe plant pests

Classification and Groups
North America: 73 families, 11,000 species

Important Families:
Gelechiidae - gelechiid moths; small to minute moths; among most injurious are Angoumois grain moth and pink bollworm

Lasiocampidae - tent caterpillars (larvae) or lappet moths (adults); some important defoliating species whose larvae form silken tents including eastern tent caterpillar and forest tent caterpillar

Noctuidae - noctuid moths; an exceptionally important group of economic pests

Pyralidae - pyralid moths; small moths which include the European corn borer

Sphingidae - hornworms (larvae) or hawk moths (adults); large moths with species of economic importance (e.g., tobacco and tomato hornworms)

Tineidae - clothes moths

Tortricide - tortricid moths; includes pests of orchards and a significant forest pest, the spruce budworm

Notes:

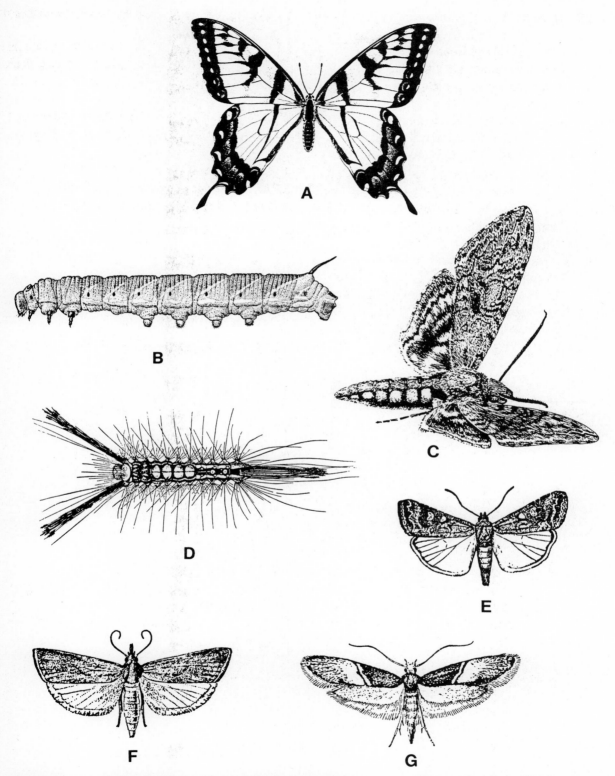

Figure 3.14   Lepidoptera. A. Monarch butterfly larva (Danaidae). B. Tomato hornworm larva (Sphingidae). C. Tomato hornworm adult (Sphingidae). D. Whitemarked tussock moth larva (Lymantriidae). E. Tiger swallowtail (Papillionidae). F. Beet armyworm (Noctuidae). G. Rice stalk borer (Pyralidae). H. Carpet moth (Tineidae). (Courtesy USDA)

Mallophaga   (*mallo* = wool; *phaga* = eat)   **chewing lice**

Description and Identification
>Metamorphosis: gradual
>Immature: like adult
>Adult:
>>Mouthparts - chewing
>>Tarsal Segments - 1 or 2
>>Size - minute (less than 5 mm)
>>Wings - none
>>Other Characteristics - head as broad or broader than thorax; small compound eyes; antennae short and may be concealed in head
>Similar Orders: Anoplura (sucking mouthparts, narrower head)

Habitat
>ectoparasites of birds and mammals

Economic Importance
>some species parasites of poultry and livestock

Classification and Groups
>North America: 7 families, 320 species

Notes:

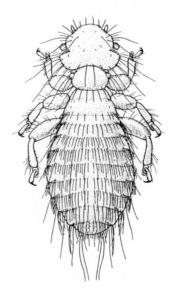

Figure 3.15   Mallophaga - chewing louse (Menoponidae). (Courtesy USDA)

Mecoptera *(meco* = long; *ptera* = wings)   **scorpionflies**

Description and Identification

Metamorphosis: complete

Immature: larvae with distinct head, thoracic legs and abdominal prolegs, most eruci-form, usually 10 or more ocelli grouped together on side of head; pupate decti-cous (exarate)

Adult:

Mouthparts - chewing, beak-like

Tarsal Segments - 5

Size - medium

Wings - 4 or none; fore and hind wings of comparable size; may be patterned

Other Characteristics - head with characteristic long face or beak; antennae long and filiform

Similar Orders: Neuroptera (head not modified into long face)

Habitat

generally on vegetation; larvae and some adults scavengers, other adults predaceous

Economic Importance

Classification and Groups

North America: 5 families, 85 species

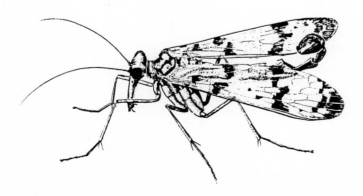

Figure 3.16   Mecoptera - scorpionfly (Panorpidae). (Courtesy USDA)

Microcoryphia   (*micro* = small; *coryphia* = head)   **jumping bristletails**

Description and Identification
        Metamorphosis: none
        Immature: like adult
        Adult:
                Mouthparts - chewing
                Tarsal Segments - 3
                Size - small
                Wings - none
                Other Characteristics - cylindrical with arched thorax; usually body with scales; large compound eyes, usually contiguous; 3 caudal filaments (appendages at end of abdomen) and styli on abdominal segments 2-9 and often on mid and hind coxae
        Similar Orders: Thysanura (lack coxal styli; separate, small compound eyes; body somewhat flattened), Diplura (only two caudal filaments)

Habitat
        debris, leaf litter, under bark, and similar sites

Economic Importance
        none

Classification and Groups
        previously included as part of Thysanura
        North America: 2 families, 20 species

Notes:

Neuroptera (*neuro* = nerve; *ptera* = wings)

alderflies, antlions, dobsonflies, fishflies, lacewings, owlflies, and snakeflies

Description and Identification

Metamorphosis: complete

Immature: varies; larvae generally campodeiform; pupae decticous exarate; larvae predaceous, aquatic in some families; many with mandibles and maxillae modified for grasping and sucking prey juices, others with chewing mouthparts (and obvious labrum)

Adult:

Mouthparts - chewing

Tarsal Segments - 5

Size - medium to large

Wings - 4; hind and fore wings of comparable size

Other Characteristics - antennae long and filiform, clavate, or pectinate; no cerci; numerous cross veins in wings

Similar Orders:

Larvae - Coleoptera (chewing mouthparts without obvious labrum or without both mandibles and maxillae modified for sucking)

Adult: Mecoptera (head modified with long face or beak), Odonata (short, setaceous antennae; wing venation different), Plecoptera (have cerci)

Habitat

on vegetation, some species aquatic

Economic Importance

important biological control agents

Classification and Groups

North America: 15 families, 350 species

Notes:

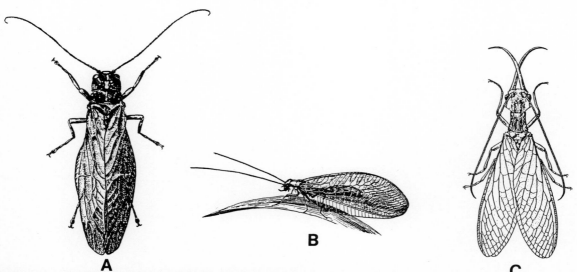

Figure 3.17   Neuroptera. A. Alderfly (Sialidae). B. Goldeneye lacewing (Chrysopidae). C. Dobsonfly (Corydalidae). (Courtesy USDA)

Odonata   (*odonata* = tooth)   **damselflies** and **dragonflies**

Description and Identification

Metamorphosis: incomplete

Immature: naiad; labium modified to catch prey, can extend up to third of body length; one group with 3 leaf-like gills at end of abdomen

Adult:

Mouthparts - chewing

Tarsal Segments - 3

Size - moderate to large

Wings - 4; membranous with numerous cross veins; fore and hind wings of approximately equal length; wings with stigma (a thickened and darkened area on the leading edge [costal margin] of the wing)

Other Characteristics - compound eyes very large; antennae short and setaceous

Similar Orders:

Immatures - Ephemeroptera (lateral abdominal gills), Plecoptera (thread-like gills on thorax)

Adults - Ephemeroptera (fore and hind wings different sizes), Neuroptera (long antennae);

Habitat

immatures aquatic, adults near water

Economic Importance

indicators of water quality

Classification and Groups

North America: 11 families, 425 species

Important Subdivisions: suborders, Anisoptera (dragonflies) and Zygoptera (damselflies)

Notes:

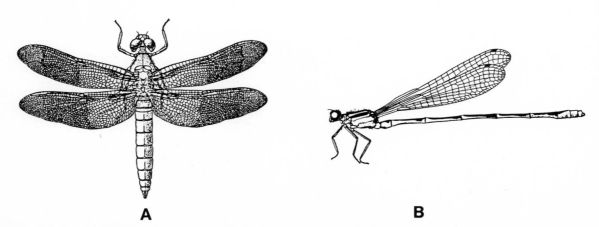

A   **B**

Figure 3.18   Odonata. A. Dragonfly (Libellulidae). B. Damselfly (Coenagrionidae). (Courtesy USDA)

Orthoptera (*ortho* = straight; *ptera* = wings)

cockroaches, crickets, grasshoppers, katydids, mantids, and walking sticks

Description and Identification

Metamorphosis: gradual

Immature: like adult, may have externally-visible wingpads

Adult:

      Mouthparts - chewing

      Tarsal Segments - 3 to 5

      Size - large

      Wings - 4 (some species with 2 or wingless); fore wings usually large, many cross veins, and thickened or hardened (called tegmina); hind wings membranous with many cross veins

      Other Characteristics - differ among groups; antennae often long and filiform; frequently long ovipositor; have 2 abdominal cerci (short appendages); various leg modifications including saltatorial hind legs (modified for jumping)

Similar Orders: Coleoptera (fore wings modified into elytra, without veins), Hemiptera and Homoptera (sucking mouthparts)

Habitat

ground and vegetation, few species in soil

Economic Importance

many species crop pests (especially grasshoppers), cockroaches important urban pests

Classification and Groups

North America: 19 families, 1,100 species

Important Subdivisions (Suborders):

(some of these groups are given ordinal status by certain authorities)

      Caelifera and Ensifera - grasshoppers, crickets, and katydids

      Phasmatodea - walking sticks

      Dictyoptera - includes superfamilies:

            Manoidea - mantids

            Blattoidea - cockroaches

Important Families:

      Acrididae - short-horned grasshoppers

      Blattelidae and Blattidae - cockroaches

      Gryllidae - crickets

      Mantidae - mantids

      Tettigoniidae - long-horned grasshoppers (includes katydids)

Notes:

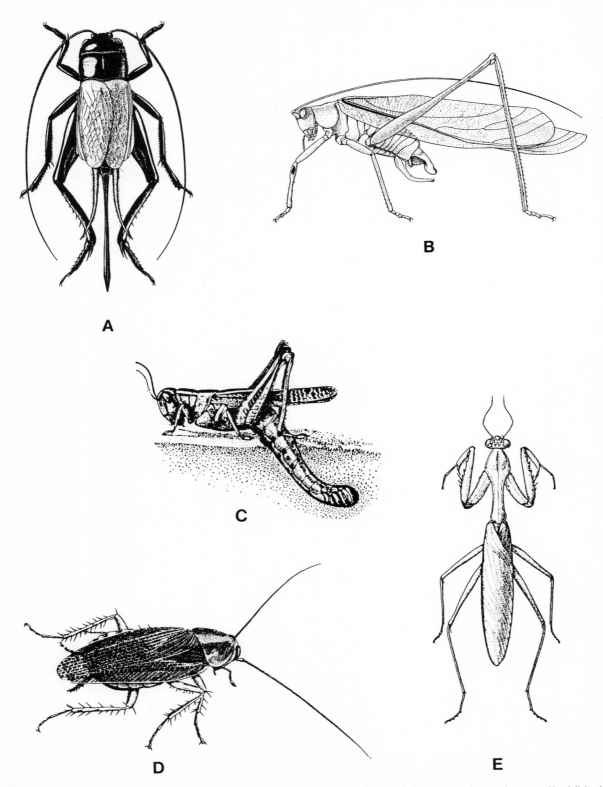

Figure 3.19 Orthoptera. A. Cricket (Gryllidae). B. Katydid (Tettigoniidae). C. Twostriped grasshopper (Acrididae). D. German cockroach (Blattellidae). E. Carolina mantid (Mantidae). (Courtesy USDA)

Plecoptera   *(pleco* = folded; *ptera* = wings) stoneflies

Description and Identification

Metamorphosis: incomplete

Immature: naiad; elongate and slightly dorsoventrally compressed; long antennae and two long cerci at end of abdomen; filamentous gills at leg bases on thorax; dorsum of thorax with plate-like sclerites; immatures may be predaceous or phytophagous

Adult:

Mouthparts - chewing

Tarsal Segments - 3

Size - medium

Wings - 4; fore wing narrow; hind wing with anal (posterior) lobe

Other Characteristics - long antennae; cerci present and often long

Similar Orders:

Immatures - Ephemeroptera (3 caudal filaments and leaf-like gills), Odonata (modified labium and no gills on thorax)

Adults - Ephemeroptera (setaceous antennae, small hind wing), Neuroptera and Trichoptera (5 segmented tarsi and lack anal lobe on hind wing)

Habitat

immatures aquatic, adults near water

Economic Importance

none, but important indicator for water quality

Classification and Groups

North America: 6 families, 400 species

Notes:

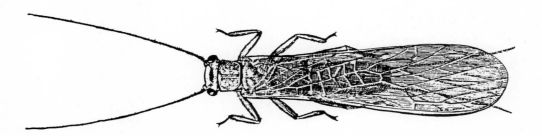

Figure 3.20   Plecoptera - common stonefly (Perlidae). (Courtesy USDA)

Protura   (*prot* = first; *ura* = tail)   **proturans**

Description and Identification
        Metamorphosis: none
        Immature: like adult; 3 molts (add abdominal segments with molts)
        Adult:
                Mouthparts - sucking
                Tarsal Segments - 1
                Size - minute
                Wings - none
                Other Characteristics - no antennae or eyes; fore legs carried forward like antennae; conical head; 12 abdominal segments in adult
        Similar Orders: other minute wingless insects (e.g., Diplura, Collembola) but Protura easily distinguished by lack of antennae, 12 abdominal segments, and lack of other characters
Habitat
        debris; feed on decaying matter and fungus
Economic Importance
        none
Classification and Groups
        North America: 3 families, 18 species
Notes:

Psocoptera (*psoco* = rub small; *ptera* = wings)

barklice and **booklice**, or **psocids**

Description and Identification

Metamorphosis: gradual

Immature: like adult

Adult:

Mouthparts - chewing

Tarsal Segments - 2 or 3

Size - minute to small (less than 5 mm)

Wings - 4 or none; hind wing smaller than fore wing; wings may be long or short

Other Characteristics - soft-bodied; face swollen or bulging; long, slender antennae; no cerci

Similar Orders: Anoplura and Mallophaga (short antennae, ectoparasites), Embioptera (modified fore tarsi), Isoptera (4 segmented tarsi, antennae short)

Habitat

in debris, under bark, occasionally in buildings; feed on dry organic matter and fungi

Economic Importance

can be a nuisance in homes when large populations develop (usually in association with fungi such as mildew)

Classification and Groups

This order has also been called Corrodentia by some workers

North America: 13 families, 280 species

Notes:

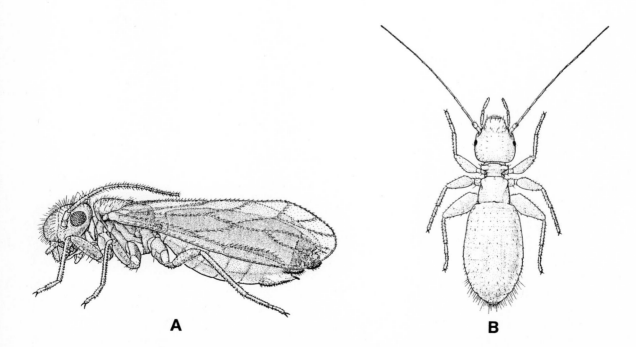

A   **B**

Figure 3.21   Pscoptera. A. Barklouse (Pseudocaeciliidae). B. Booklouse (Liposcelidae). (Courtesy USDA)

Siphonaptera   (*siphon* = a tube; *aptera* = wingless)   **fleas**

Description and Identification

Metamorphosis: complete

Immature: larvae white, cylindrical, eucephalic, legless, with long setae on thorax and abdomen and 2 small hooks on posterior abdominal segment, feed on various organic matter (often from host); pupae adecticous exarate in silken cocoon

Adult:

Mouthparts - sucking

Tarsal Segments - 5

Size - minute

Wings - none

Other Characteristics - ectoparasites of birds and mammals; body generally oval having a small head and being laterally compressed and heavily sclerotized, with numerous bristles and setae; legs long and modified for jumping

Similar Orders: Anoplura and Mallophaga (dorsoventrally compressed)

Habitat

ectoparasites of birds and mammals

Economic Importance

pests on human pets; medically important as vectors of plague and various other diseases

Classification and Groups

North America: 7 families, 250 species

Notes:

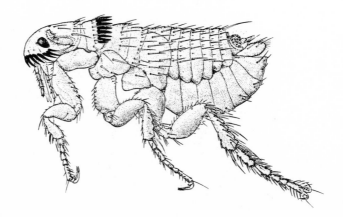

Figure 3.22   Siphonaptera - cat flea (Pulicidae). (Courtesy USDA)

Strepsiptera   (strepsi = twisted; ptera = wings)   twisted-winged parasites

Description and Identification
Metamorphosis: complete

Immature: internal parasites of insects; 1st stage larvae with well-developed eyes and legs, subsequent larval stages legless and worm-like; pupae adecticous exarate

Adult:
Mouthparts - vestigial

Tarsal Segments - 2 to 5

Size - minute

Wings - female wingless; male with 2 hind wings and fore wings modified into club-like structures (similar to halteres)

Other Characteristics - most parasitic with female wingless and legless, living entirely in host; end of parasite may protrude between host abdominal segments

Similar Orders: Homoptera (without modification of fore wings), Hymenoptera (difficult to distinguish larvae, adults not parasites)

Habitat
insect parasites

Economic Importance
none (too uncommon to be a major biological control agent)

Classification and Groups
Included in Coleoptera as family (Stylopidae) by some workers

North America: 4 families, 60 species

Notes:

Thysanoptera   (*thysano* = fringe; *ptera* = wings)   thrips

Description and Identification

    Metamorphosis: gradual
    Immature: like adult
    Adult:

        Mouthparts - rasping-sucking
        Tarsal Segments - 1 or 2
        Size - minute
        Wings - 4 or none; wings long and narrow, without veins but fringed with long
            hairs
        Other Characteristics - slender with short legs; short antennae (6-9 segments);
            mouthparts form conical beak on the ventral base of the head, mouth-
            parts are asymmetric
    Similar Orders - none

Habitat

    in debris or on vegetation

Economic Importance

    a number of species are pests of cultivated plants, besides feeding injury some can
    vector plant diseases

Classification and Groups

    The common name, thrips, is the correct singular and plural form; proper usage is to
    speak of a thrips or many thrips.
    North America: 5 families, 600 species

Notes:

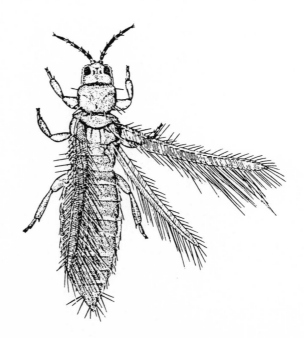

Figure 3.23   Thysanoptera - flower thrips (Thripidae). (Courtesy USDA)

Thysanura   (*thysan* = fringe; *ura* = tail)   **bristletails**

Description and Identification
Metamorphosis: none
Immature: like adult
Adult:

Mouthparts - chewing
Tarsal Segments - 3 to 5
Size - small
Wings - none
Other Characteristics - elongate and somewhat flattened; small, separated compound eyes; body usually with scales and long antennae; 3 caudal filaments (appendages at end of abdomen); abdominal styli on segments 2-9, 7-9, or 8-9, but no coxal styli

Similar Orders: Microcoryphia (large, converged compound eyes; cylindrical body with arched thorax), Diplura (only 2 caudal filaments)

Habitat
debris, under bark, in buildings (some species cool, damp conditions, others warmer)

Economic Importance
Two species (silverfish and firebrat) household pests, feed on starchy or sweet materials (e.g., bookbinding)

Classification and Groups
North America: 3 families, 25 species
Important Families:

Lepismatidae - includes silverfish and firebrat

Notes:

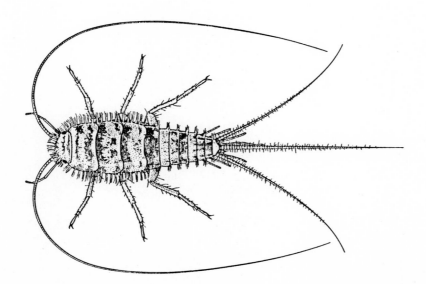

Figure 3.24   Thysanura - firebrat (Lepismatidae).   (Courtesy USDA)

Trichoptera   (*tricho* = hair; *ptera* = wings)   **caddisflies**

Description and Identification
Metamorphosis: complete

Immature: aquatic; larvae eruciform with pair of hooks towards posterior of abdomen, often in cases (caddises) made of stones, twigs, etc.; pupae decticous (exarate)

Adult:

Mouthparts - chewing (reduced)

Tarsal Segments - 5

Size - small to medium

Wings - 4; hairy

Other Characteristics - antennae long and filiform; well-developed palps but reduced mandibles

Similar Orders: Lepidoptera (wings scaled and usually with coiled proboscis)

Habitat
aquatic or near water

Economic Importance
no pests, may be an important indicator species of water quality

Classification and Groups
North America: 17 families, 975 species

Notes:

Figure 3.25   Trichoptera - caddisfly (Phryganeidae). (Courtesy USDA)

Zoraptera (*zor* = pure; *aptera* = wingless)

Description and Identification
    Metamorphosis: gradual
    Immature: like adult
    Adult:
        Mouthparts - chewing
        Tarsal Segments - 2
        Size - minute
        Wings - 4 or none; membranous with hind wing smaller than fore wing
        Other Characteristics - antennae filiform or moniliform; cerci short and 1 segmented with long bristle
    Similar Orders: Embioptera, Isoptera, and Psocoptera (all with different tarsi, antennae, and cerci)

Habitat
    in debris, under bark, sawdust; primarily feed on fungi

Economic Importance
    none

Classification and Groups
    North America: 1 family, 2 species

Notes:

4 INSECT GROWTH AND DEVELOPMENT

Immature and adult insects frequently are very different in appearance and habits. For instance, the dragonfly hovers and glides at the shore's edge, capturing airborne-prey, whereas its aquatic immature, or naiad, dwells under water, waiting for a daphnia or other prey to swim by; a butterfly flits through a meadow feeding on the nectar of flowers whereas its larval form, a caterpillar, lives on a plant where it feeds on the leaves.

Because immature insects are common and abundant, and because they often represent the damaging stages of economically-important insect pests, it is important to be able to recognize and identify the immature forms of a variety of insects. Eggs, juveniles, nymphs, larvae, pupae, and adults all have unique structures and functions; therefore, the better we can know and understand these, the better we can properly identify an insect and then evaluate the pest management problems it presents.

Unfortunately, insect collectors frequently overlook immature insects because of the difficulties in preserving and identifying them. Because many immatures are soft-bodied and have a high fluid content, they tend to decay and putrefy rapidly, and they cannot, generally speaking, be mounted on pins. Instead, they must be mounted on microscope slides or placed in vials filled with 75-80% ethyl alcohol following treatment with boiling water or special killing solutions (see Appendix 4.1). However, even with proper preservation there is a tendency for many preserved immature insects to change color and shape over time. It is therefore most desirable, when possible, to identify an immature insect while it is still alive or freshly killed.

OVIPOSITION

Egg laying, or **oviposition**, begins the insect **life cycle**, or the sequence of biological events that occurs during the lifetime of an individual insect. Oviposition takes three forms, depending on the stage deposited. Most insects exhibit **oviparity**, where embryonic development proceeds within the egg after it has been deposited. The egg is surrounded by a protective outer shell, or **chorion**, and the immature insect hatches from the egg some time after oviposition. Some insects exhibit **ovoviviparity**. Ovoviviparity occurs in certain species in the orders Ephemeroptera, Orthoptera, Psocoptera, Homoptera, Thysanoptera, Lepidoptera, Coleoptera, and, especially, Diptera. In these insects, eggs with a chorion are formed, but they are retained within the female's body until embryonic development is nearly completed. Emergence of the immature insects from the eggs usually occurs when the eggs are still in the female's body or immediately upon oviposition. A specialized form of birth occurs in insects exhibiting **viviparity**. In viviparous insects, the immature insects, after completing embryonic development, are actually nurtured for some additional time within the body of the female. They are they released from her in some advanced state of development. The tsetse fly, *Glossina* spp., for example, gives birth to mature larvae which become pupae soon after they are deposited. Similarly, many aphid species are viviparous. Essentially, viviparity may be defined as the "oviposition" of active immature insects.

ECLOSION

The postembryonic growth and development of insects generally begins with the hatching, or **eclosion**, of the egg. The typical insect egg is covered by a two-layered chorion that provides support and protection from water loss or uptake. The outer layer of the chorion frequently is elaborately sculptured. The sculpturing reflects the form of the follicle cells of the ovarioles that secrete the chorion. Chorionic sculpturing often is a useful character for identifying insect eggs.

The strength of the chorion can make eclosion a challenging process, and insects have evolved some ingenious solutions that allow them to escape the confines of the egg. Most insects swallow air or embryonic fluids thereby increasing the turgidity and volume of the body. Further pressure can be exerted against the inside of the chorion when waves of muscular contraction are employed. The chorion may rupture irregularly or along pre-existing lines of weakness. In the latter case, a little cap, or **operculum** is present that the insect pops open like a lid. Some insects simply chew their way out of the chorion, and some others secrete enzymes which assist the insect in breaking out of the egg. Many insects hatch with the aid of specialized structures, typically on the head, known as **egg bursters**. Egg bursters are variable in form, occurring as spines, teeth, or structures everted by hemolymph pressure. They generally function in piercing, cutting, or pushing against the chorion.

GROWTH OF IMMATURES

Once hatched, the immature insect begins to feed and grow. Because the insect cuticle, or **exoskeleton**, cannot stretch or expand appreciably to accommodate a growing and changing insect, the cuticle periodically must be shed and replaced. Growth in insects is thus punctuated by a series of molts. **Molting** may be defined as the periodic digestion of most of the old cuticle, secretion of a new, larger cuticle, and shedding of the undigested portion of the old cuticle. The shed cuticle is called an **exuvium** and the actual shedding of the exuvium is sometimes referred to as **ecdysis**. Although brief, the period between ecdysis and the hardening of the newly-secreted cuticle leaves the insect in a vulnerable condition in which it may easily succumb to predators or adverse environmental conditions.

Epidermal cells are the workhorses of the molting cycle. They are the targets for the molting hormone, **ecdysone**, which is released from the prothoracic glands to trigger the molting processes. At the onset of molting, the epidermal cells show activity increases and morphological changes. The epidermal cells separate from the cuticle (a condition called **apolysis**), secrete a **molting fluid** containing various digestive enzymes that degrade the old cuticle, reabsorb the digested constituents of the old cuticle, and lay down the multilayered new cuticle. Once the secretion of the new cuticle is completed, ecdysis may proceed.

As it matures, the typical immature insect progresses through a series of molts and ecdyses, generally increasing in size with each. The form assumed by an insect between molts is known as an **instar** and the time period between molts is referred to as a **stadium**. The term **stage** is frequently used when referring to the insect's developmental status (e.g., larval stage) or when describing developmental status after specific molts (e.g., a third-stage larva is that which follows the second molt). When an insect has secreted its new cuticle but has not yet escaped from the exuvium comprised of the prior cuticle, it is said to be in the **pharate** condition. The final developmental stage is the adult, or **imago**; it is during this stage that sexual maturity usuually is achieved and functional wings (if they are to be present) are attained.

Primitive insects usually exhibit more molts than do the advanced species (e.g., some Ephemeroptera have up to 25 molts while many Diptera have only four). More specialized insects, such as parasites and parasitoids, tend to have fewer molts.

Growth in insects, as in all multicellular organisms, occurs as the result of an increase in the number of cells in the body and, in many instances, an increase in cell size. We can define **growth** as a lasting increase in weight, volume, and linear dimensions. It may be more or less gradual and continuous or it may be discontinuous with periods of growth alternating with static intervals of little or no growth.

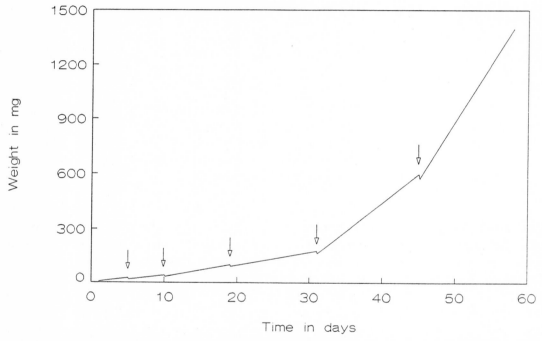

Figure 4.1 The pattern of weight increase of female *Locusta*. The time of the molts is indicated by arrows. (Redrawn from Clark 1957)

Typically, insect weight increases progressively through each immature instar and then falls slightly at molting due to the loss of part of the cuticle at ecdysis and to a loss of water as feeding ceases through a portion of the molt (Fig. 4.1). Increases in insect length and surface area are generally regarded as discontinuous due to the relative inflexibility of the exoskeleton. Growth of hardened, or **sclerotized**, portions of the body must occur in a series of steps, each step corresponding to a molt. In certain types of insects, the predictability and regularity of the increase in size at each molt has permitted the formulation of **growth laws**. The most important of such principles is **Dyar's law**. It is based on the assumption that the linear dimensions of an insect increase geometrically at each molt, the ratio being constant for a given stage of a particular species. Thus, plotting the logarithm of a measurement of some linear dimension (e.g., femur length or head capsule width) against instar number yields a straight line. In insects where this law applies, the relationship may be exploited in determining the number of instars of an insect; this is because if the dimensions of the final instar plus the ratio of increase between any two successive instars are known, interpolation can allow deduction of: (1) the dimensions of the other instars and (2) the number of instars, found by counting the number of points on the generated curve (Fig. 4.2).

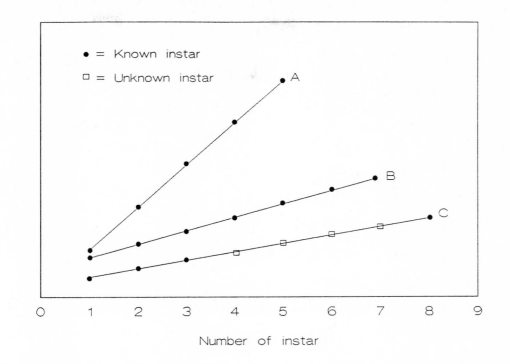

Figure 4.2 Graph depicting Dyar's Law as it might apply to three hypothetical insect species. A, B, and C are curves generalized by plotting the log of a linear measurement against the number of the instar. For C, given the known instar information, one can determine the number of instars and the linear dimension associated with each instar. (Redrawn from Romoser 1973)

Dyar's law has some practical applications in terms of pest management. For instance, in order to assess the damage potential of some lepidopteran pest populations it is necessary to know the size and number of larvae present and the number of remaining insect stages. The relationship between head capsule or thoracic shield widths and larval stage (Fig. 4.3) has thereby proven useful in rapidly evaluating the stages comprising a given population and thus, in providing information necessary for making proper management decisions.

Dyar's law implies and assumes that insect growth is **harmonic** or **isogonic** (i.e., that all parts of the insect and insect's body as a whole increase in size by the same ratio during each molt). While harmonic growth is fortuitously followed in some insects, the majority of insects exhibit **heterogonic** or **allometric** growth. That is, some parts of the insect body develop at rates different than other parts. Heterogonic growth has been described by the equation:

$$y = bx^k$$

where y = the length of a body part, x = the length of the body, b = a constant (the factorial coefficient), and k = the growth coefficient, a ratio of the specific components of the growth rate of y to the specific components of the growth rate of x. In practice, this equation may be used in determining if growth in a particular insect is harmonic or heterogonic. X and y values are plotted against one another on log-log paper. If a straight line is obtained, the slope of that line, or k, is determined. If k is greater or less than 1, then growth may be regarded as heterogonic. If k is equal to 1, then growth may be considered harmonic.

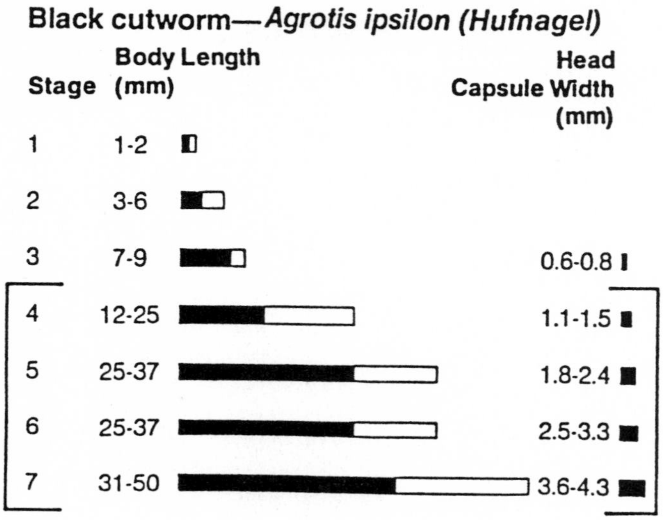

Black cutworm—*Agrotis ipsilon (Hufnagel)*

Stage	Body Length (mm)		Head Capsule Width (mm)	
1	1-2	▯		
2	3-6	▮▯		
3	7-9	▮▮▯	0.6-0.8	▮
4	12-25	▮▮▮▯	1.1-1.5	▪
5	25-37	▮▮▮▮▯	1.8-2.4	▪
6	25-37	▮▮▮▮▯	2.5-3.3	▪
7	31-50	▮▮▮▮▮▯	3.6-4.3	▪

Figure 4.3   The head capsule measurements and approximate body lengths corresponding to larval stages are depicted for the black cutworm, *Agrotis ipsilon*. For body length measurements, the shaded region corresponds to the minimum length of a stage, and the unshaded portion represents the potential size range for the stage. The stages within the brackets are considered the most destructive. (Courtesy Iowa Cooperative Extension Service)

Factors Influencing Growth

The rates at which insects grow is influenced by time and the environment, with temperature being of particular importance. Because insects are **poikilothermic** (i.e., have body temperatures which fluctuate with the environmental temperatures), their development, reproduction, and other life processes are intimately influenced by temperature. Therefore, temperature may be used, in conjunction with a measurement of time, in predicting various events in an insect's life cycle. **Degree days** are measurements which combine time and temperature to yield predictions regarding insect development. Sometimes degree days are referred to as heat units, thermal units, or growing degree days. In any case, degree days represent the number of degrees above some minimum temperature necessary for growth multiplied by time in days. Ten degrees above the minimum for 5 days represents 50 degree days (10 x 5) just as does 2 degrees above the minimum for 25 days (25 x 2). Both instances represent the same amount of physiological time, i. e., an insect would have grown the same amount under either condition. Some total number of degree days are associated with different developmental stages. For example, the alfalfa weevil requires 200 degree days for eggs to hatch and an additional 300 to grow from newly ecolosed larva to pupa.

The first requirement for calculating degree days is to recognize that growth only occurs within a range of temperatures. The minimum temperature below which no growth occurs is called the **minimum developmental threshold** or the minimum threshold. Growth will increase with higher temperatures up to a maximum temperature called the **maximum**

developmental threshold. These thresholds are determined experimentally and are different for each insect species. Although a minimum threshold is required for insect development, maximum thresholds often are not included since many insects rarely experience temperatures near the maximum threshold during development. The degree day concept and degree day calculation is discussed in more detail in Chapter 11.

Metamorphosis

At hatching, an insect is often very different in morphology from the adult insect. The developmental process by which the immature stage transforms into the adult stage is called **metamorphosis** which literally means "change in form".

Insects are frequently categorized according to the extent of the changes in their metamorphosis. **Ametabolous** insects (Fig. 4.4) have no metamorphosis, and the adult form differs from the immature only in size and by possessing fully-developed gonads and external genitalia. Apterygotes such as silverfish (Thysanura) and springtails (Collembola) exhibit ametabolous development; the immature form in these insects is called a **juvenile**. Adults and juveniles generally live in the same environments and have the same requirements for survival.

The pterygote insects are categorized as either **hemimetabolous** or **holometabolous**. In hemimetabolous insects, metamorphosis often is described as either "gradual" or "incomplete", with immatures assuming forms which become, with each molt, increasingly similar to the adult form. Insects with the gradual form of hemimetabolous development (Fig. 4.5) have an immature form known as a **nymph**. Nymphs typically resemble adults but lack fully-formed external genitalia and wings. Later instars often have externally-visible beginnings of wings, called **wing pads**. Both nymphs and adults are usually found in the same habitat, feeding on the same foods. Insects demonstrating the gradual type of hemimetabolous development include the grasshoppers and cockroaches (Orthoptera), the true bugs (Hemiptera), and the planthoppers, treehoppers, and aphids (Homoptera).

Insects exhibiting the incomplete type of hemimetabolous development (Fig. 4.6) are also characterized by a lack of functional external genitalia and the presence, in later instars, of external wing pads. The immatures, called **naiads**, are found in or near water and frequently have modified structures, such as tracheal gills, which permit aquatic respiration. Adults are usually found in somewhat different environments than the naiads and feed upon a different selection of foods. Mayflies (Ephemeroptera), dragonflies (Odonata), and stoneflies (Plecoptera) are all characterized by this type of metamorphic development.

Holometabolous development, or complete metamorphosis (Fig. 4.7), occurs in more highly-evolved insects including the true flies (Diptera), bees and wasps (Hymenoptera), beetles (Coleoptera), and moths and butterflies (Lepidoptera). In addition to an egg and adult, there is a **larva** and a **pupa**. With few exceptions, the larval stage is very different from the adult: it exhibits a different body form, lacks compound eyes, has reduced antennae, and lacks external evidence of wing development. In most instances, the larva and adult occur in separate environments and feed differently, thereby eliminating much of the competition between stages. Typically, the larval stage is regarded as specialized for feeding while the adult stage is designed for dispersal and reproduction. The pupal stage is usually a quiescent stage during which the histological and morphological changes which permit a larva to become an adult insect occur. In many instances, such as in the moths, the pupa is relatively immobile and is enclosed in a silken cocoon or other protective enclosure. In other instances, such as the mosquitoes, the pupa is an active, mobile stage.

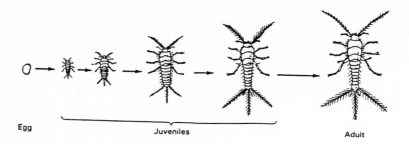

Figure 4.4 The no-metamorphosis model of development represented by a silverfish, *Lepisma saccharina*. (Redrawn from Little 1972)

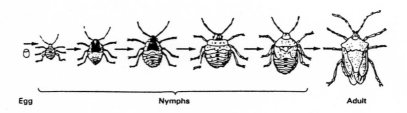

Figure 4.5 The gradual-metamorphosis model of development represented by a stink bug (Hemiptera: Pentatomidae). (Redrawn from Little 1972)

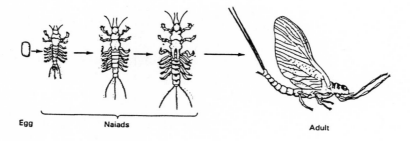

Figure 4.6 The incomplete-metamorphosis model of development represented by a mayfly (Ephemeroptera). (Redrawn from Little 1972)

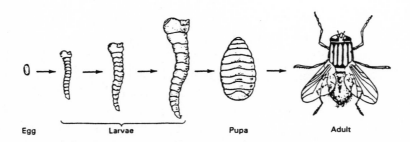

Figure 4.7 The complete-metamorphosis model of development represented by a house fly, *Musca domestica*. (Redrawn from Little 1972)

In most holometabolous insects, the larval instars resemble each other rather closely in appearance and habits. However, some holometabolous insects pass through one or more larval instars which are quite different from the others . This phenomenon is termed **hypermetamorphosis** (Fig. 4.8) and has been described in a number of insect orders and families, especially those exhibiting highly-specialized forms of parasitism and inquilinism (i.e., living as a guest or occupant in the nests of social insects).

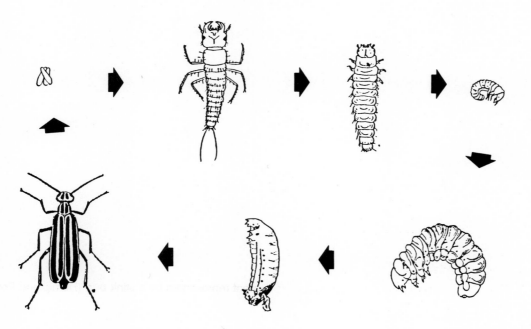

Figure 4.8   The hypermetamorphic form of holometabolous development represented by the striped blister beetle, *Epicauta vittata*. (Courtesy USDA)

Control of Metamorphosis

Metamorphosis is under the control of the insect neuro-endocrine systems. There are three principle organs involved in the regulation of metamorphosis in all types of insects. These are (1) groups of **neurosecretory cells** in the brain, (2) small, white glandular organs posterior to the brain called the **corpora allata**, and (3) glandular tissue in the thorax called the **prothoracic glands**. The neurosecretory cells, in response to various internal or external stimuli, produce **brain hormone**. This hormone acts as a link between the environment and other hormones and is stored in small structures called **corpora cardiaca** just posterior to the corpora allata. Brain hormone is released from the corpora cardiaca to the hemolymph. One target of the brain hormone is the prothoracic glands, which secrete the molting hormone, ecdysone (Fig. 4.9A). As mentioned earlier, ecdysone stimulates epidermal cells to initiate molting processes. Brain hormone also stimulates the corpora allata to produce **juvenile hormone** (Fig. 4.9B). While ecdysone concentration determines whether or not a molt will occur, juvenile hormone concentration determines whether a larva molts to another larger larva or proceeds with the development of adult characters. The presence of high amounts of juvenile hormone in the hemolymph suppresses the expression of adult characteristics while an absence of the hormone permits the development of adult structures. The principle targets for juvenile hormone are the **imaginal disks**. These are groups of cells with the potential for growth and differentiation into adult structures. Juvenile hormone has the ability to inhibit imaginal disk growth. However, in the absence of juvenile hormone, imaginal disks may develop and the metabolism of larval tissues, with concurrent genesis of adult tissues, may proceed.

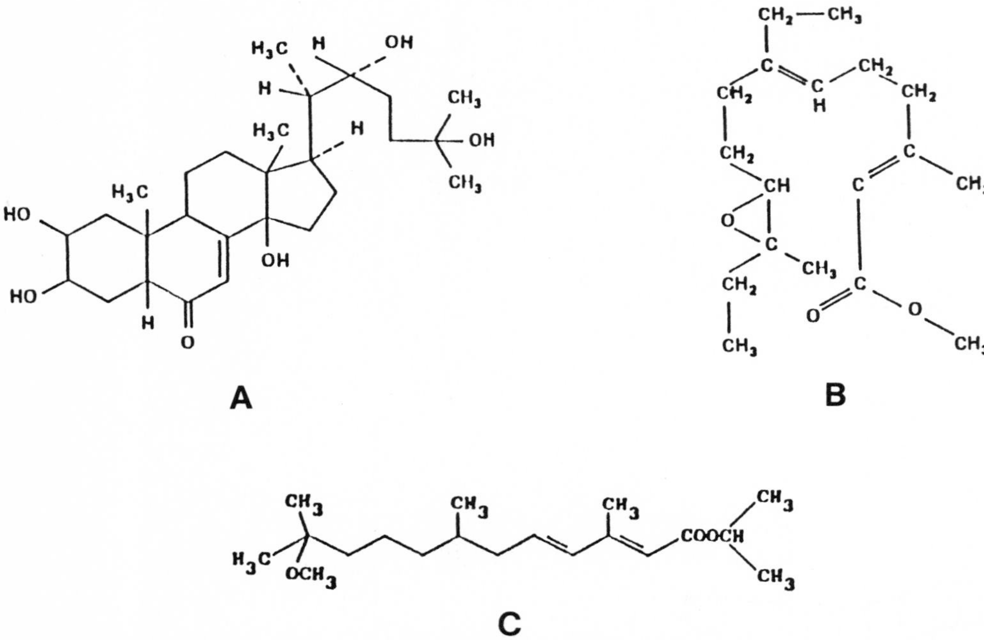

Figure 4.9 Chemical structures of (A) ecdysone, (B) juvenile hormone, and (C) methoprene.

The hormonal control of metamorphosis has been exploited in the development of new insect-management tools. The so-called **third-generation insecticides** include synthetic juvenile hormone analogs such as methoprene (Fig. 4.9C), which can severely impair normal insect development, and other growth-influencing materials such as the chitin-synthesis inhibitors. **Chitin** is a nitrogenous polysacaccharide (n-acetyl-D-glucosamine) that gives the insect exoskeleton its strength.

Types of Larvae

The immature forms of holometabolous insects assume a variety of forms. For descriptive purposes, it is often convenient, however, to group the larvae into several categories based on general appearance. **Campodeiform** larvae (Fig. 4.10A) are named for their resemblance to the dipluran genus *Campodea*. They have dorso-ventrally flattened bodies, long legs, and, in many cases, long antennae and cerci. Campodeiform larvae are observed in the Neuroptera, Trichoptera, and some Coleoptera.

Carabiform larvae (Fig. 4.10B) are named for their resemblance to larvae of the coleopteran family Carabidae, the ground beetles. These larvae are similar to the campodeiform type but have shorter legs and cerci. This larval type characterizes several beetles besides the carabids.

Elateriform larvae (Fig. 4.10C) are named for their resemblance to the larvae ("wireworms") of the click beetles of the Elateridae family of coleopterans. These larvae are typified

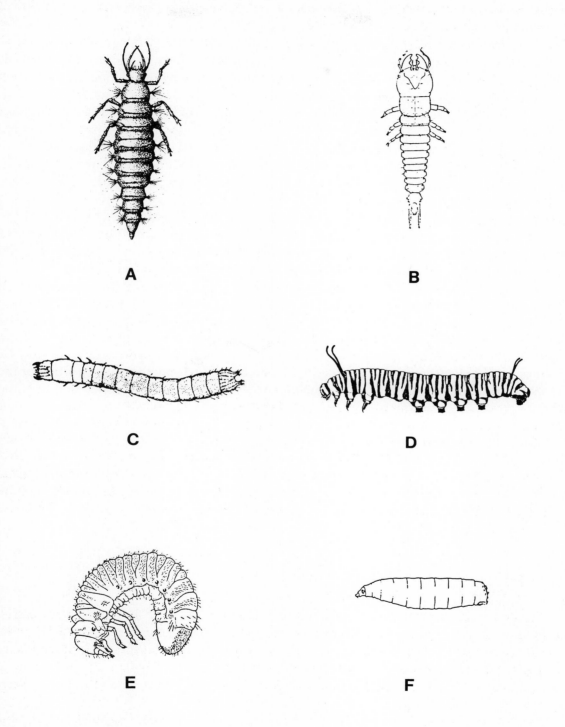

Figure 4.10   Various forms of insect larvae. A. Campodeiform. B. Carabiform. C. Elateriform. D. Eruciform. E. Scarabaeiform. F. Vermiform (A Courtesy Colorado Cooperative Extension Service; B,D,E,F Courtesy USDA; C Courtesy Illinois Cooperative Extension Service)

by cylindrical, smooth, and hardened bodies. There are distinct heads and short legs. Elateriform larvae are found in several other Coleoptera families including Tenebrionidae, the darkling beetles or mealworms.

Eruciform larvae (Fig. 4.10D) are typical "caterpillars". They occur in the Lepidoptera and some Hymenoptera and Mecoptera. Eruciform larvae have cylindrical bodies, well-developed heads, short antennae, short thoracic legs, and often, stubby abdominal legs called **prolegs**. The number and distribution of prolegs are characters used in identification. The prolegs of lepidopterans exhibit tiny hooks, or **crochets**, on their lower surfaces. Crochet arrangement is also a diagnostic character for many lepidopteran larvae. The pattern of hairs, or **setae**, on the body surface of the eruciform larva is of additional use in identification.

Platyform larvae get their name from the Greek for broad and flat. They may be legless or have very short thoracic legs. This larval form is found in certain Lepidoptera, Coleoptera, and Diptera but is relatively uncommon.

Scarabaeiform larvae (Fig. 4.10E) are named for their resemblance to larvae of the Coleoptera family Scarabaeidae, the scarab beetles. These larvae are often called "grubs". They have heavy, cylindrical bodies which are frequently curled into a C-shape, well-developed heads, and thoracic legs. Scarabaeiform larvae occur in a number of the Coleoptera.

Vermiform larvae (Fig. 4.10F) are worm-like in appearance and are often referred to as "maggots". Vermiform larvae typify a number of Diptera, and also occur in some Hymenoptera, Lepidoptera, Coleoptera, and the Siphonaptera. They lack legs but may be adorned with various spines and tubercles. The head capsule or the mouthparts often are the only sclerotized parts of the body. With regards to the head, the vermiform larva may be described as **eucephalic** (the head is well-defined), **hemicephalic** (only the apical portion of the head is sclerotized and well-defined), or **acephalic** (the head is not well-defined and only the mouthhooks, or cephalopharyngeal armature, are exposed and sclerotized). Another character frequently used in the identification of dipteran vermiform larvae is the number and distribution of spiracles on the body.

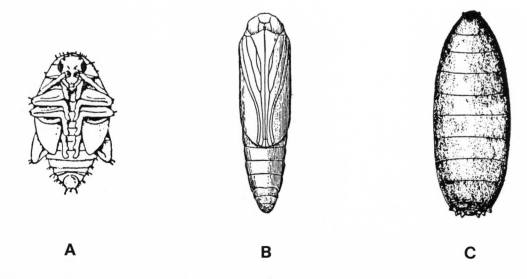

A   **B**   **C**

Figure 4.11   Forms of insect pupae. A. Exarate. B. Obtect. C. Coarctate. (Courtesy USDA)

Types of Pupae

The pupae of holometabolous insects also may be categorized by morphological characteristics. Pupae which possess powerful, sclerotized, articulated mandibles used in escaping from the cocoon or pupal cell are called **decticous**. Neuroptera, Mecoptera, most Trichoptera, a few Lepidoptera, and a very few Hymenoptera have decticous pupae. All decticous pupae are **exarate**, or have appendages that are free from the body (Fig. 4.11A).

A second pupal type is termed **adecticous**. Adecticous pupae have nonarticulated mandibles that are not movable and not used in exiting the cocoon or pupal cell. Adecticous pupae may be either exarate or **obtect**. Obtect adecticous pupae (Fig. 4.11B) have appendages which are immobile, glued down to the body by a secretion produced at the larva-pupa molt. Obtect pupae more often are sclerotized than are exarate pupae, and they are frequently covered by some type of cocoon. All higher Lepidoptera, some Coleoptera, many Diptera, and a few Hymenoptera are adecticous and obtect. Exarate adecticous pupae, in which the appendages are free of any secondary attachment to the body, occur in the Siphonaptera, Strepsiptera, most Coleoptera, most Hymenoptera, and many Diptera. Some adecticous exarate pupae are encased in the hardened cuticle (termed **puparium**) of the penultimate larval instar. This pupa is often described as **coarctate** (Fig. 11C). Coarctate pupae occur in some of the dipterans.

OBJECTIVES

1) Understand how insects grow and change and know how brain hormone, ecdysone, and juvenile hormone influence growth and development.

2) Know and understand the sequence of events that typifies the life cycle of an ametabolous, hemimetabolous, and holometabolous insect.

3) Be able to categorize larval specimens, on the basis of morphological characteristics, as: campodeiform, carabiform, elateriform, eruciform, platyform, scarabaeiform, and vermiform.

4) Understand the differences between decticous and adecticous pupae and obtect and exarate pupae.

5) Additional objectives (at instructor's discretion):

OPTIONAL DISPLAYS AND EXERCISES

1) Examine the displayed egg, larval, and pupal specimens.

2) Examine the displayed examples of ametabolous, hemimetabolous, and holometabolous insects. Recognize and be aware of the morphological and ecological differences between the immatures and adults.

3) Investigate the role of temperature in insect development by rearing, to pupae and adults, insect larvae of the house fly, wax moth, black cutworm, or other species under two or more temperature regimes. Be sure to begin with eggs or larvae of uniform age and control all other environmental factors.

4) Determine if the growth of a specific insect is heterogonic or harmonic. Measure some linear dimensions (x) of the insect of interest (e.g., femur length of cockroaches, head capsule width of lepidopteran larvae). Also measure the overall body lengths (y) of the insects. Plot x and y against one another on log-log paper. If the process of measuring fits the heterogonic growth equation, $y = bx^k$, then a straight line should be obtained. Determine the slope (k) of the line. If k is less than or greater than 1, then growth is heterogonic; if k = 1, then growth is harmonic.

5) Additional displays or exercises (at instructor's discretion):

TERMS

acephalic	eclosion	isogonic	oviposition
adecticous	egg bursters	juvenile	ovoviviparity
allometric growth	elateriform	juvenile hormone	pharate
ametabolous	epidermal cells	life cycle	platyform
apolysis	eruciform	maximum	poikilothermy
brain hormone	eucephalic	developmental	prolegs
campodeiform	exarate	threshold	prothoracic glands
carabiform	exoskeleton	metamorphosis	puparium
chitin	exuvium	minimum	scarabaeiform
chorion	growth	developmental	sclerotization
coarctate	growth laws	threshold	setae
corpora allata	harmonic growth	molting	stadium
corpora cardiaca	hemicephalic	molting fluid	stage
crochets	heterogonic growth	naiad	third-generation
degree days	holometabolous	neurosecretory cells	insecticide
decticous	hypermetamorphosis	nymph	vermiform
Dyar's law	imaginal disks	obtect	viviparity
ecdysis	imago	operculum	wing pads
ecdysone	instar	oviparity	

DISCUSSION AND STUDY QUESTIONS

1) The counting of insect eggs and egg masses has been used as a means of estimating potential populations of specific insect species in various crops. What are the advantages and disadvantages of sampling eggs for population estimation?

2) What do you see as the advantages and disadvantages of oviparity, viviparity, and ovoviviparity in insects?

3) Synthetic analogs of juvenile hormone show much promise as "third generation insecticides". What are the benefits of using such compounds for insect management? What are the drawbacks?

4) Most insects exhibiting hypermetamorphosis have a highly mobile and very active first stage larva called a "triungulin". What is the purpose of the triungulin?

5) What advantages for survival were conferred upon insects through the evolution of holometabolous development? How might pest management approaches differ for hemimetabolous vs. holometabolous pest species?

BIBLIOGRAPHY

Pedigo, L. P. 1989. Entomology and pest management. Macmillan Pub. Co., New York, NY
- Chapter 4. The insect life cycle.

Blum, H. S. 1985. Fundamentals of insect physiology. John Wiley and Sons, New York, NY
- a multiauthor text with excellent chapters on the integument and hormonal control of development.

Chapman, R. F. 1982. The insects: structure and function. 3rd. ed. Harvard Univ. Press, Cambridge, MA
- a comprehensive and accurate text on insect morphology and physiology.

Chu, H. F. 1949. How to know the immature insects. W. C. Brown Co., Dubuque, IA
- a useful little book that includes keys to immature insects. Descriptions of groups and keys are not as comprehensive as in other references. We have had difficulty with some of the keys.

Higley, L. G., L. P. Pedigo, and K. R. Ostlie. 1986. DEGDAY: A program for calculating degree-days, and assumptions behind the degree-day approach. Environ. Entomol. 15: 999-1016
- a useful summary of the degree-day approach. Includes information on assumptions associated with degree days and on various calculation methods. A computer program is included.

Hinton, H. E. 1981. The biology of insect eggs (in three volumes). Pergamon Press, New York, NY
- a complete coverage of the insect egg and its biology, physiology, and morphology. Contains a large number of photomicrographs illustrating chorionic sculpturing and other microstructure of eggs of many insect species.

Jungreis, A. M. 1979. Physiology of moulting in insects. Adv. Insect Physiol. 14: 109-183
- though slightly dated, this is a thorough summary of the molting process.

Locke, M. 1974. The structure and function of the integument in insects. in Rockstein, M. (ed.), The physiology of Insecta. 2nd ed. vol. VI. Academic Press, New York, NY
- a complete discussion of the composition, properties, functions, and formation of the insect integument.

Peterson, A. 1962. Larvae of insects. Part I and II (in separate volumes), Edwards Brothers, Inc., Ann Arbor, MI
- for decades the major reference to larval insects. Includes keys and numerous illustrations. It is now replaced by the newly-released "Immature Insects".

Pruess, K. P. 1983. Day-degree methods for pest management. Environ. Entomol. 12: 613-619
- a discussion of degree day calculation methods as they pertain to pest management.

Raabe, M. 1982. Insect neurohormones. Plenum Press, New York, NY
- a complete review of work on the roles and functions of neurohormones in insects including their involvement in insect growth and development.

Schwahn, F. E. 1988. Insect morphogenesis. Karger, New York, NY
- although this book focuses largely on embryonic development of insects, it also contains discussions on metamorphosis, oviposition, and a valuable chapter on insect culture and egg collection.

Stehr, F. W. (ed.) 1987. Immature insects. Vol. 1. Kendall Pub. Co., Dubuque, IA
- an important new reference to immatures that will include keys to order and families for all North American groups. This volume treats all groups except the Hemiptera, Homoptera, Neuroptera, Coleoptera, Strepsiptera, Diptera, and Siphonaptera which will be covered in a forthcoming second volume.

Wigglesworth, V. B. 1972. The principles of insect physiology. 7th ed. Methuen, London
- a classic text covering many aspects of insect physiology

FIGURE CITATIONS

Clark, K. U. 1957. On the increase in linear size during growth in *Locusta migratoria* L. Proc. Royal Entomol. Soc. London A, 32:35-39.

Little, V. A. 1972. General and applied entomology. 3rd ed. Harper and Row Pub. Co., Inc. New York, NY

Romoser, W. S. 1973. The science of entomology. Macmillan Pub. Co., Inc., New York, NY

APPENDIX 4.1 - KILLING SOLUTIONS FOR IMMATURE INSECTS

Immature insects should be killed by immersion in boiling water for one to five minutes or by submersion in one of the following killing solutions. After one to several hours in the killing solution, immatures should be transferred to 75-80% ethyl alcohol for storage.

K.A.A.D. Solution

kerosene	1 part
95% ethyl alcohol	8 parts
glacial acetic acid	1 part
dioxane	1 part

This solution works well for Diptera and Lepidoptera larvae, Hymenoptera larvae and pupae, Neuroptera larvae, and Coleoptera larvae. It is not very good for heavily sclerotized larvae such as wireworms or for naiads. After treatment in this solution, transfer immatures to 95% ethyl alcohol.

X.A.A.D. Solution

xylene	1 part
refined isopropyl alcohol	6 parts
glacial acetic acid	5 parts
dioxane	4 parts

This solution works well for Lepidoptera and Coleoptera larvae. After treatment in this solution, transfer immatures to 75-80% ethyl alcohol.

X.A. Solution

xylene	1 part
95% ethyl alcohol	1 part

This solution works well for Lepidoptera, Coleoptera, and some Hymenoptera larvae. After treatment in this solution, transfer immatures to 75-80% ethyl alcohol.

SECTION II.   INSECT PESTS

The following six chapters focus on important insect pests of the United States. These insect species are arranged into different groupings based on the system they injure or affect; specifically these groupings are: Chapter 5: Agronomic Pests; Chapter 6: Horticultural Pests; Chapter 7: Forest, Shade Tree and Related Pests; Chapter 8: Medical and Veterinary Pests; Chapter 9: Stored Product Pests; and Chapter 10: Urban Pests. In addition to learning about the insect pests in these systems, you will need to understand the systems themselves so you can relate insect pest management to other production or pest management requirements.

The four steps in curing an insect pest problem are: (1) to identify the insect, (2) to quantify the numbers or effect of the insect, (3) to evaluate if management is warranted, and (4) to undertake the appropriate management tactic. In focusing on individual insects much of the emphasis is on insect identification, but you also should obtain insights into management options and constraints for different insects and commodities.

Because much of the practical focus of these chapters is on insect identification, you may wish to review material from earlier chapters. Chapter 1 on identification and preservation, Chapter 2 on morphology, and Chapter 3 on systematics and insect orders all provide background information you need for the following sections on insect pests.

Included in the insect pest chapters are lists of major insect pests by commodities. Commodities were chosen based on factors such as acreage or value of production. Choosing the most important insects in these systems is less objective. Unfortunately, most statistics are unavailable for acreages injured by or insecticides used against a given insect species. Although pests that always occur in damaging numbers are obviously important, many occasional or sporadic pests occur in many commodities and can be equally significant. For this book we have tried to identify species that are most commonly injurious, but hundreds of additional species could have been included. The importance of an insect pest is dynamic, varying between geographic locations and through time. You or your instructor may recognize more important species for certain situations in your area.

Although you learn insect pest management as a single subject, all pest management is conducted in the context of a production system. In agriculture, production systems are centered around individual commodities. Even insect pests we don't associate with a particular crop or commodity will require different management in different systems (for example, cockroach management in food preparation facilities is different from cockroach management in households). You need to understand these different production systems, so you can recognize how insect pest management fits in a given system. Many aspects of production can influence crop susceptibility to insect injury and the choice of management tactics. Practices such as rotations, planting dates, tillage patterns, and variety selection can be used to reduce or eliminate some insect problems, but this use may conflict with other considerations. For example, fall tillage can be used to manage insects overwintering in plant residue, but it greatly increases the likelihood of soil erosion. Likewise, aspects of crop physiology are important for understanding the relationship between a crop and its insect pests. For example, soybean yields are stable over a wide range of plant populations, therefore, stand reductions from insect injury are not likely to be as injurious as stand reductions in other crops. Even utilization of a crop may be important for pest management. For instance, we can tolerate much less sunflower moth injury on sunflower grown for human consumption or birdseed than on sunflower grown for oil. Once you are familiar with various production systems and their associated pests, you can begin to make inferences about new situations and look for analogies between different systems.

Information on individual insects is useful both for identification and management. Making identification notes will help you learn to sight-identify a species, but also remember that situation and population indices will aid in field identifications. Origin of an insect may indicate management options; for example, introduced species may be good candidates for biological control. Similarly, the type of injury, e.g., defoliation vs. pod feeding, will greatly influence our management options. In terms of economic losses, crops can tolerate less direct injury (injury to yield-producing organs) than indirect injury. And when an insect vectors an important plant pathogen, the presence of even a single insect vector may result in unacceptable losses on the crop. Through understanding an insect's seasonal cycle, we may be able to identify vulnerable points that can be exploited. For example, many insects overwinter in plant residue or debris, and are extremely susceptible to environmental disruption (such as tillage, sanitation, or residue destruction). Thus, you need to know more than just the identifying characteristics for insect pests.

These chapters will be most useful to you if you look for similarities and differences across commodities and insects. Try to recognize fundamental characteristics of different insect groups and to identify common pest management procedures for different commodities. By developing a broader appreciation you will be better prepared to meet actual pest management situations.

These chapters provide an introduction to different pest management systems, but this information is limited. Many injurious species are not discussed and information on management tactics for specific situations is not included. Consequently, you need to learn how to find and use reference information for insect pest management. Unfortunately, no single compendium is available for identifying economically important insects, although various references do provide identification keys, pictures, and information on many pests. Practical information on management procedures is available from the Cooperative Extension Service. You need to be able to find and use these references. For some insects information on life history or management practices may be unavailable. It is in these situations that your broader understanding of insect groups and production systems will be of most value in making rational pest management decisions.

GENERAL BIBLIOGRAPHY

Davidson, R. H., and W. F. Lyon. 1986. Insect pests of farm, garden, and orchard. 8th ed. John Wiley and Sons, New York, NY

Pfadt, R. E., editor. 1985. Fundamentals of applied entomology. 4th ed. Macmillan Pub. Co., New York, NY
    - both useful introductions to economic entomology with emphasis on insects in various commodities; earlier editions may include information on different pests. Although less comprehensive than Metcalf, Flint, and Metcalf, the information in these texts is far more current.

Entomological Society of America. Environmental Entomology and Journal of Economic

Entomology.
- these two scientific journals, particularly JEE, are the major outlets for research information on economically important insects in the US

Fichter, G. S. 1966. Insect pests. Golden Press, New York, NY
- a useful little guide with excellent illustrations to many insect pests.

Frankie, G. W., and Koehler, C. S., eds. 1983. Urban entomology: interdisciplinary perspectives. Praeger Publishers, New York, NY
- a thought provoking volume with chapters on many areas of urban pest management. Unlike many multiauthored books, this is well-written and edited. It includes chapters relating to medical, urban, stored product, shade tree, and ornamental pests. Most chapters focus on principles and theory rather than details of pest management for specific insects or situations.

Harris, T. W. 1862. A treatise on some of the insects injurious to vegetation. Orange, Judd, and Co., New York, NY
- This was the first, or at least one of the first, books ever written on applied entomology and insect pests. Although it is primarily of historical interest, much information on insect life histories is still of value. Originally published in the 1850's, the 1862 edition includes color plates. Copies can be found in many university libraries.

Metcalf, C. L., W. P. Flint, and R. L. Metcalf. 1962. Destructive and useful insects. 4th ed. McGraw-Hill Book Co., New York, NY
- for decades this was the standard reference on economic insects in the US. Although dated, out of print, and missing various newer pests, it remains the most comprehensive reference available.

Wilson, M. C. and others. 1980. Practical insect pest management series. 6 vol., Waveland Press, Inc., Prospect Heights, IL
- six volumes on insect pest management in a variety of systems. Some emphasis on pest management in the Midwest and considerable information on pest management practices. Overall these are very useful references. Individual titles are: Fundamentals of Applied Entomology; Insects of Livestock and Agronomic Crops; Insects of Vegetables and Fruit; Insects of Ornamental Plants; Insects of Man's Household and Health; and Insect Pests of Forests.

5 AGRONOMIC PESTS

Although we may think of civilization as having depended on the creation of government, religion, ethics, science, or art, the undeniable fact is that ultimately our society is based on food, and more specifically, on grain. The domestication of natural grasses, that are now the small grains, and the subsequent development of agricultural practices made civilization possible. And as has been the case for thousands of years, our present societies still depend on the production of grain as a bulwark against starvation and chaos.

Table 5.1   Major agronomic crops in the United States, 1980 & 1985. (Data from USDA Agricultural Statistics, 1981 and USDA Agricultural Statistics, 1986, U.S. Government Printing Office, Washington, D.C.; na = data not available.)

Crop	Harvested (millions of acres)		Value (billions of $)	
	1980	1985	1980	1985
barley	7.2	11.5	1.0	1.2
corn:				
grain	70.1	75.1	21.7	21.3
forage & silage	9.9	7.4	na	na
cotton	13.2	10.2	4.1	3.5
dry bean	1.8	1.6	0.7	0.4
hay (ca. 40% alfalfa)	59.4	60.6	8.1	9.7
oats	8.6	8.1	0.8	0.6
peanuts	1.4	1.5	0.6	0.9
rice	3.3	2.5	1.7	1.1
sorghum:				
grain	12.7	16.7	1.8	2.4
forage & silage	2.1	1.3	na	na
soybean	67.9	61.6	13.8	10.8
sugarbeet	1.2	1.1	1.1	na
sugarcane	0.7	0.8	1.0	na
sunflower	3.7	2.8	0.4	0.3
wheat	70.8	64.7	9.4	7.7

The **agronomic**, or field and forage, crops are without question the most important agricultural products in the US and the world, in terms of acreage, value of production, and human consumption. Agronomic crops include fiber crops such as cotton, legumes, such as alfalfa and soybean, and the grasses, such as corn and the small grains. Major agronomic crops in the US are listed in Table 5.1, with the most important being corn, soybean, wheat, and hay (alfalfa and grasses). Among the primary uses of agronomic crops are human food, livestock forage and feed, fiber, and oil.

Insects are important pests in many agronomic production systems. In some instances insects may actually limit production; in others, losses from insects and management costs do not so much limit production as reduce profitability. In either case, insect pest management is a significant consideration in crop production.

128

PEST MANAGEMENT FOR AGRONOMIC PESTS

Insect pest management depends on a consideration of economic and environmental criteria in making management decisions. Environmental factors most often are accommodated through the choice of a management tactic, whereas economic factors involve crop physiology, the value of crop production, and the type of insect injury. A central theme in insect pest management is the notion of tolerating some insects, that controlling all insect injury is not economically justified. Tolerating insect injury is particularly important for understanding how insect pest management operates for agronomic crops.

Tolerating Injury

Because of mechanization, high-yielding cultivars, fertilization, and other production practices, the agronomic crops generally have low production costs. Low production costs and large levels of production contribute to relatively low crop values per unit area as compared to other crops such as vegetables. When crop values per acre are low, it takes more insect injury to produce economic losses than with a high cash value crop. Consequently, in most instances more insect injury can be tolerated on agronomic crops than on higher cash value crops.

This relationship between crop values and tolerating insect injury is particularly evident with perennial grasses and forages. The native perennial grasses of rangelands in the American West have an extremely low value per acre, so insect pest management programs most often are associated with extremely damaging situations, such as grasshopper outbreaks. On other grasslands with higher production and greater value, e.g., irrigated grasslands or pastures, less insect injury is tolerable. For hay production, such as of grasses or alfalfa, even less insect injury can be tolerated. However, in all these situations greater numbers of insects and levels of insect injury may be acceptable than with the production of a grain crop.

Besides these differences between crops, our ability to tolerate injury can differ within a single plant species. For example, alfalfa grown for hay may be less intensively managed and less valuable than alfalfa grown for a specific purpose, such as fodder for dairy cattle. Thus, alfalfa for dairy cattle is likely to be more valuable and cannot tolerate as much insect injury. A more extreme example is when alfalfa, or indeed any commodity, is grown for seed. Seed crops are much more valuable than regular crops, and therefore cannot tolerate as much injury. In fact, pest management on crops produced for seed may be vastly different than on the same crop grown for other purposes. On alfalfa, for instance, the important pest species are entirely different than on alfalfa grown for hay. In this manual we have not focused on pests in seed-production systems, but you should recognize that the important pests and pest management programs are likely to be different with seed production.

Beyond value of production, both crop physiology and the type of insect injury are important considerations in how many insects and how much injury is tolerable. Both of these factors relate to how much reduction in yield is associated with a given number of insects in a given crop. Some agronomic crops can tolerate substantial insect injury without excessive yield loss. For example, soybean is injured by an array of insect species, but soybean plants can compensate or tolerate much of this injury without large yield reductions. As a comparison, consider injury by the seedcorn maggot which will destroy germinating seeds of both corn and soybean. Destruction of 10% of germinating soybean seeds is unlikely to produce a yield loss, but the loss of 10% of germinating corn seeds will cause a measurable loss in yield.

Types of Injury

The type of injury also has an obvious relationship to yield reductions. Those insects inflicting **direct injury**, injury to yield producing organs, have a much greater effect on yield than do insects causing **indirect injury**, injury to non-yield producing organs. Thus, more indirect injury is tolerable than direct injury. Many agronomic insect pests produce direct injury. For example, although defoliation can substantially reduce cotton yields, all the major cotton insect pests attack the squares and bolls, the yield-producing organs. Similarly, in sunflower one of the most injurious insects is the sunflower moth, whose larvae feed in sunflower heads on seeds. One type of indirect injury that is comparable to direct injury is when the indirect injury kills the plant. Agronomic insects such as cutworms, rice water weevil, seedcorn maggot, or wireworms that destroy seeds or developing plants cannot be tolerated as readily as insects producing less severe effects. Insects that are **vectors** (organisms capable of transmitting pathogens) of plant pathogens may not be acceptable even at extremely low levels, particularly if the pathogen is highly virulent.

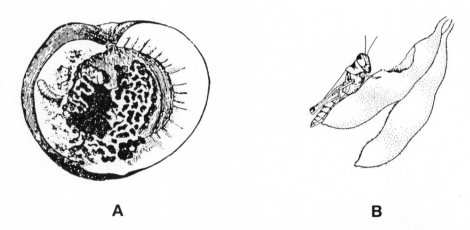

A   **B**

Figure 5.1   Examples of direct injury. A. Mediterranean fruit fly larva feeding on apple. B. Grasshopper feeding on soybean pod. (A Courtesy USDA; B Courtesy Illinois Cooperative Extension Service)

The formal mechanism for considering how well a crop can tolerate a certain number of insects (and their associated injury) is through the **economic injury level** (or EIL), defined as the number of insects or level of injury at which management is economically justified. Based on this definition, the **higher** the EIL, the greater the number of insects or level of injury that can be tolerated in a crop. We will consider economic injury levels in more detail in Chapter 11, but they provide a useful scale for comparing pest management in different crops and for different insects. Generally, EILs are much higher for insects in agronomic crops than in horticultural crops. Among the agronomic crops, EILs are higher for insects in grasslands and forages than in grain crops. Additionally, insects in crops grown for seed will have lower EILs than insects in the same crops grown for grain or fodder. Moreover, generally EILs are higher for insects producing indirect injury than for insects producing direct injury.

Management Options

Tolerating injury is an important consideration in pest management programs, especially in choosing management tactics. Many of our management tactics require that some injury is tolerated. When we cannot tolerate much injury (i.e., when an insect has a low EIL), our choice of tactics may be limited, often to insecticides. For example, we mentioned that all major cotton pests produce direct injury. These insects have relatively low EILs and consequently, insect pest management in cotton is very dependent on insecticides.

However, for most agronomic insects various management options are available. Biological control has been used successfully with some agronomic insects, but management may be limited to certain regions. (Because agronomic crops and their associated pests are widely distributed in various climatic regions, it is often difficult to establish natural enemies in all of these areas.) In general, host plant resistance, cultural and chemical tactics are used most often with agronomic insects. Host plant resistance has been especially important as a management option for aphids and the Hessian fly. Cultural techniques such as fall tillage, modified planting or harvest dates, rotations, and destruction of plant residues can be important options in managing agronomic insects. An important limitation to some of these approaches, however, is that use of some of these techniques may conflict with other production considerations, such as soil erosion control. Insecticides also are important in managing pests of many crops. Although pest management programs are designed to reduce insecticide use, many situations in agronomic crops continue to rely on chemical tactics. Soil-inhabiting insects are a particular problem in this regard. Frequently, insects in soil cannot be sampled easily, they may destroy seeds or seedlings, and insect management may be impossible after planting. Consequently, many soil-inhabiting insects are managed through the use of prophylactic insecticide applications. This dilemma regarding soil insect management is important for various agronomic insects, as well as for insects in many other commodities.

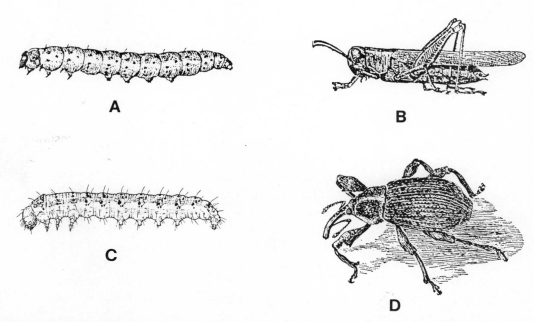

A

B

C

D

Figure 5.2   Some important agronomic pests. A. European corn borer. B. Grasshopper. C. Corn earworm. D. Boll weevil. (Courtesy USDA)

Another important consideration for pest management in agronomic crops is the diversity of insect pests that may cause injury. Because agronomic crops are grown over such large acreages, even occasional pests that attack only a fraction of the available crop may injure hundreds of thousands of acres. Locally, occasional and sporadic insect pests can cause substantial, even devastating, losses. In some systems, especially in forage production which often is less intensively managed, injury from occasional pests may be overlooked, resulting in excessive loss. One challenge for insect pest management on agronomic crops is to recognize these occasional pests and to have appropriate management programs for these insects.

OBJECTIVES

1) Understand how insect pest management in agronomic crops differs from management in other commodities.

2) Recognize common pest management tactics for agronomic insect pests.

3) Know basic production characteristics and identify their implications for insect pest management in the following commodities: alfalfa, corn, cotton, grasses (forage, pasture, and range), soybean, and wheat.

4) Know major features of life histories, and be able to sight identify the following insect groups: armyworms and cutworms, aphids, grasshoppers, leafhoppers, and wireworms.

5) Know major features of life histories and be able to sight identify the following insect species: alfalfa weevil, boll weevil, chinch bug, corn earworm, corn rootworm (northern and western), European corn borer, and two-spotted spider mite.

6) Additional objectives (at instructor's discretion):

OPTIONAL DISPLAYS AND EXERCISES

1) Select two commodities not in objective 3 (your choice) and learn production practices and insect pests (including identification).

2) Determine recommended management practices for major insect pests of an agronomic crop grown in your area.

3) Examine the displayed seed pests for different commodities.

4) Examine the displayed groups of insects (which include pests and nonpests) and identify the insect pest.

5) Examine the displayed examples of injury by agronomic insects.

6) Additional displays or exercises (at instructor's discretion):

TERMS

agronomic   direct injury   economic injury   indirect injury
                                           level

MAJOR AGRONOMIC PESTS

alfalfa weevil	*Hypera postica*	Coleoptera: Curculionidae
armyworms, especially:		
armyworm	*Pseudaletia unipuncta*	Lepidoptera: Noctuidae
beet armyworm	*Spodoptera exigua*	Lepidoptera: Noctuidae
fall armyworm	*Spodoptera frugiperda*	Lepidoptera: Noctuidae
bean leaf beetle	*Cerotoma trifurcata*	Coleoptera: Chrysomelidae
beet leafhopper	*Circulifer tenellus*	Homoptera: Cicadellidae
beet webworm	*Loxostege sticticalis*	Lepidoptera: Pyralidae
black grass bugs	*Labops* and *Irbisia* spp.	Hemiptera: Miridae
blue alfalfa aphid	*Acyrthosiphon kondoi*	Homoptera: Aphididae
boll weevil	*Anthonomus grandis grandis*	Coleoptera: Curculionidae
chinch bug	*Blissus leucopterus leucopterus*	Hemiptera: Lygaeidae
corn earworm	*Heliothis zea*	Lepidoptera: Noctuidae
= bollworm/tomato fruitworm		
corn rootworms:		
northern corn rootworm	*Diabrotica barberi*	Coleoptera: Chrysomelidae
western corn rootworm	*Diabrotica virgifera virgifera*	Coleoptera: Chrysomelidae
cotton fleahopper	*Pseudatomoscelis seriatus*	Hemiptera: Miridae
cutworms, especially:		
black cutworm	*Agrotis ipsilon*	Lepidoptera: Noctuidae
dingy cutworm	*Feltia ducens*	Lepidoptera: Noctuidae
variegated cutworm	*Peridroma saucia*	Lepidoptera: Noctuidae
European corn borer	*Ostrinia nubilalis*	Lepidoptera: Pyralidae
grasshoppers, especially:		
bigheaded grasshopper	*Aulocara elliotti*	Orthoptera: Acrididae
clearwinged grasshopper	*Camnula pellucida*	Orthoptera: Acrididae
differential grasshopper	*Melanoplus differentialis*	Orthoptera: Acrididae
migratory grasshopper	*Melanoplus sanguinipes*	Orthoptera: Acrididae
redlegged grasshopper	*Melanoplus femurrubrum*	Orthoptera: Acrididae

twostriped grasshopper	*Melanoplus bivittatus*	Orthoptera: Acrididae
green cloverworm	*Plathypena scabra*	Lepidoptera: Noctuidae
greenbug	*Schizaphis graminum*	Homoptera: Aphididae
Hessian fly	*Mayetiola destructor*	Diptera: Cecidomyiidae
lesser cornstalk borer	*Elasmopalpus lignosellus*	Lepidoptera: Pyralidae
lygus bugs:		
pale legume bug	*Lygus elisus*	Hemiptera: Miridae
tarnished plant bug	*Lygus lineolaris*	Hemiptera: Miridae
western lygus bug	*Lygus hesperus*	Hemiptera: Miridae
Mexican bean beetle	*Epilachna varivestis*	Coleoptera: Coccinellidae
Mormon cricket	*Anabrus simplex*	Orthoptera: Tettigoniidae
pea aphid	*Acyrthosiphon pisum*	Homoptera: Aphididae
pink bollworm	*Pectinophora gossypiella*	Lepidoptera: Gelechiidae
potato leafhopper	*Empoasca fabae*	Homoptera: Cicadellidae
rice stink bug	*Oebalus pugnax*	Hemiptera: Pentatomidae
rice water weevil	*Lissorhoptrus oryzophilus*	Coleoptera: Curculionidae
seedcorn maggot	*Delia platura*	Diptera: Anthomyiidae
sorghum midge	*Contarinia sorghicola*	Diptera: Cecidomyiidae
southern green stink bug	*Nezara viridula*	Hemiptera: Pentatomidae
soybean looper	*Pseudoplusia includens*	Lepidoptera: Noctuidae
spotted alfalfa aphid	*Therioaphis maculata*	Homoptera: Aphididae
sugarcane beetle	*Euetheola humilis rugiceps*	Coleoptera: Scarabaeidae
sugarcane borer	*Diatraea saccharalis*	Lepidoptera: Pyralidae
sunflower moth	*Homoeosoma electellum*	Lepidoptera: Pyralidae
threecornered alfalfa hopper	*Spissistilus festinus*	Homoptera: Membracidae
tobacco budworm	*Heliothis virescens*	Lepidoptera: Noctuidae
twospotted spider mite	*Tetranychus urticae*	Acari: Tetranychidae
velvetbean caterpillar	*Anticarsia gemmatalis*	Lepidoptera: Noctuidae
wheat jointworm	*Tetramesa tritici*	Hymenoptera: Eurytomidae
wheat stem sawfly	*Cephus cinctus*	Hymenoptera: Cephidae
wireworms	Elateridae spp.	Coleoptera: Elateridae

INSECT PESTS BY COMMODITY

Alfalfa

    alfalfa weevil, blue alfalfa aphid, pea aphid, potato leafhopper, and spotted alfalfa aphid

Barley

    chinch bug, greenbug, and grasshoppers

Beans

    bean leaf beetle, cutworms, lygus bugs, Mexican bean beetle, potato leaf hopper, and seedcorn maggot

Corn (maize)

    chinch bug, corn earworm, corn rootworms (northern and western), cutworms, and European corn borer

Cotton

    irrigated far West (Arizona and California) - lygus bugs, pink bollworm; arid Southwest

(New Mexico and Texas) - boll weevil and cotton fleahopper; Midsouth and Southeast - boll weevil, lygus bugs; important secondary pests - bollworm and budworm

Grasses (hay, pasture, and range)

hay and pasture - chinch bug and grasshoppers; range - black grass bugs and grasshoppers

Oats

aphids, grasshoppers, and leafhoppers

Peanut

corn earworm, lesser cornstalk borer, and southern corn rootworm

Rice

rice stink bug and rice water weevil

Sorghum

chinch bug, corn earworm, and sorghum midge

Soybean

North (sporadically) - bean leaf beetle, green cloverworm, and twospotted spider mite; South - southern green stink bug, threecornered alfalfa hopper, defoliator complex (corn earworm, green cloverworm, soybean looper, and velvetbean caterpillar)

Sugarbeet

beet leafhopper, beet webworm, and grasshoppers

Sugarcane

sugarcane beetle, sugarcane borer, and wireworms

Sunflower

cutworms and sunflower moth

Wheat

armyworms, Hessian fly, wheat jointworm, and wheat stem sawfly

DISCUSSION AND STUDY QUESTIONS

1) Why might armyworms and cutworms be a problem in fields with poor perennial weed control?

2) Will economic injury levels for European corn borers attacking corn grown for grain be smaller, larger, or the same as for European corn borers attacking corn grown for silage? Explain.

3) Which features of aphid life histories contribute to outbreaks?

4) How might pest management tactics differ for insect pests that are native versus insect pests that are introduced?

5) Upland rice is grown without flooding; how might pest management problems differ from regular rice production?

6) How might pest management tactics differ between native and introduced insect pests? Between migratorya and non-migratory insect pests?

7) You notice large gouges and tears in crop leaves in field margins; what is probably causing the injury?

8) Why is destruction of crop residue an important management practice for many insect pests?

BIBILIOGRAPHY

Commodities

American Society of Agronomy. Monographs (various editors and titles). Madison, WI
- the ASA publishes a series of monographs on production and physiology of major agronomic crops. These are probably the best single volume references on individual crops. Titles include: Alfalfa Science and Technology; Barley; Clover Science and Technology; Corn and Corn Improvement; Cotton; Soybean: Improvement, Production and Uses; Sunflower Science and Technology; and Tall Fescue

Heath, M. E., R. F. Barnes, and D. S. Metcalfe, editors. 1985. Forages: the science of grassland agriculture. Iowa State University Press, Ames, IA
- a thorough treatment of forage grasses and legumes, including coverage of minor forages; minimal discussion of pests

Martin, J. H., W. H. Leonard, and D. L. Stamp. 1976. Principles of field crop production. 3rd ed. Macmillan Pub. Co., New York, NY
Metcalf, D. S., and D. M. Elkins. 1980. Crop production. 4th ed. Macmillan Pub. Co., New York, NY
- both of these books are excellent summaries of agronomic crops and their production in the US; Martin, Leonard, and Stamp provide more information on pests.

Diseases

American Phytopathological Society. Compendia of Plant Diseases (various editors and titles). St. Paul, MN
- the APS publishes a series of disease compendia that describe all biotic and abiotic diseases affecting major crops (agronomic, horticultural, and silvicultural) in the US. Each compendium provides a thorough treatment of diseases in a given commodity and includes numerous color illustrations. Compendia are periodically updated, and new titles are being added. Our only criticism is that the compendia do not give any indication of the relative importance of the various diseases. (Like insect pests, it is difficult to evaluate the importance of most diseases because the significance of any disease will vary through time and between regions.) If only comparable compendia were available for insect pests! Compendia currently are available for: alfalfa, barley, corn, cotton, oats, peanut, sorghum, soybean, and wheat.

Insects

Besides references listed in the introduction to insect pests section, most references to agronomic insect pest management and to individual insect pests are from USDA, Cooperative Extension Service, or state university publications. These may treat insect pests and pest management on a national, regional, or state-by-state basis and may be very difficult to locate. Listed below are a number of publications of this type that we find useful.

Capinera, J. L. 1986. Field key for identification of caterpillars found on field and vegetable crops in Colorado. Cooperative Extension Service, Colorado State University. Bulletin 535A
- a well-illustrated key to various injurious caterpillars.

Edwards, C. R., editor. Alfalfa: a guide to production and integrated pest management in the Midwest. North Central Regional Extension Publication 113
- a comprehensive guide to production and integrated pest management on alfalfa.

Hantsbarger, W. M. 1979. Grasshoppers in Colorado. Cooperative Extension Service, Colorado State University. Bulletin 502A
- some information on identification, substantial information on life histories.

Hewitt, G. B., E. W. Huddleston, R. J. Lavigne, D. N. Ueckert, and J. G. Watts. 1974. Rangeland entomology. Soc. for Rangeland Management, Denver, CO
- a useful summary of important insect pests and management procedures for rangelands.

University of California, Statewide IPM Project. IPM manuals (various contribu tors and titles). Davis, CA
- this is a superb series on integrated pest management in various commod ities, focused on California or western states. Beautifully illustrated with profuse color photographs and clear line drawings. These manuals include considerable information on insects, diseases, and weeds including substantial discussion of management practices, but they frequently emphasize insect pest management. The emphasis on California somewhat limits the usefulness of the manuals elsewhere, particularly with respect to information on pest management practices. Among agronomic crops, manuals are available for: alfalfa, cotton, and rice.

Knutson, H., S. G. F. Smith, and M. H. Blust. 1983. Grasshoppers: identifying species of economic importance. Cooperative Extension Service, Kansas State University. Bulletin S-21
- illustrated color key to major grasshopper species.

Kogan, M., and D. E. Kuhlman. 1982. Soybean insects: identification and management in Illinois. Agricultural Experiment Station, University of Ill inois. Bulletin 773
- a superior treatment of soybean insects and management in the Midwest. It includes color photographs of soybean insects, substantial information on individual pests, and keys to adults, larvae, and soybean injury.

McBride, D. K., D. D. Kopp, C. Y. Oseto, and J. D. Busacca. 1983. Insect pest management for sunflower. Cooperative Extension Service, North Dakota State University. Bulletin 28

Oseto, C. Y. 1981. Key to larvae attacking sunflowers in North Dakota. Cooperative Extension Service, North Dakota State University. Circular E-708

Weinzierl, R., D. D. Kopp, and C. Y. Oseto. 1981. Key to adults of insects commonly attacking sunflowers in North Dakota. Cooperative Extension Service, North Dakota State University. Circular E-707

- these three publications provide well-illustrated keys and information on individual sunflower insect pests.

Riley, T. J., and A. J. Keaster. A pictorial field key to wireworms attacking corn in the Midwest. USDA/SEA Extension Integrated Pest Management Program, University of Missouri. Bulletin MP 517

- keys, color plates, and information on important wireworm species.

Rings, R. W., and G. J. Musick. 1976. A pictorial field key to the armyworms and cutworms attacking corn in the north central states. Ohio Research and Development Center, Wooster, OH. Research Circular 221

Rings, R. W. 1977. An illustrated field key to common cutworm, armyworms, and looper moths in the north central states. Ohio Research and Development Center, Wooster, OH. Research Circular 227

Rings, R. W. 1977. A pictorial field key to the armyworms and cutworms attacking vegetables in the north central states. Ohio Research and Development Center, Wooster, OH. Research Circular 231

- excellent identification guides to larval and adult armyworms and cutworms. These guides include keys, line drawings, and black and white photographs.

Smith, R. H., editor. Cotton pest management in the southern United States. Alabama Cooperative Extension Service, Auburn University. Circular ANR 194

Wilde, G., and K. O. Bell. 1980. Identifying caterpillars in field crops. Agricultural Experiment Station, Kansas State University. Bulletin 632.

- includes illustrated keys and full color photographs of caterpillars on alfalfa, corn, sorghum, soybean, and wheat in the Midwest

Womack, H., J. C. French, F. A. Johnson, S. S. Thompson, and C. W. Swann. 1981. Peanut pest management in the Southeast. Cooperative Extension Service, University of Georgia. Bulletin 850

6 HORTICULTURAL PESTS

Horticulture, the science of growing vegetables, fruit, flowers, and ornamental plants, encompasses a tremendous diversity of plant species and a correspondingly large diversity of insect pests. The most important groups of horticultural crops are the vegetables and fruits but additional groups include mushrooms, nuts, turfgrass, flowers, foliage plants, and woody ornamentals. (Because insect pests of woody ornamentals are similar to pests of other woody plants, we discuss them with shade tree and forest pests in Chapter 7.) The large number of different horticultural species, each with associated cultural practices and pests, poses particular challenges for pest management.

Horticultural versus Agronomic Crops

Horticultural crops differ from agronomic crops in many respects. Because grain is a staple of human diets, huge production of a relatively few grain crops is necessary to meet human needs. Although some fruits and vegetables are staples in parts of the world, generally the horticultural crops are not primary food crops and therefore rank below agronomic crops in terms of acreage and total value of production. For example, the total value of US production for agronomic crops in 1980 was ca. $66.2 billion versus ca. $12.3 billion in 1980 for vegetables, mushrooms, fruit, and nuts. Similarly, many individual agronomic crops have annual values in excess of $1 billion, but only a few horticultural commodities reach this level.

Although the total value of horticultural production is less than that of agronomic crops, on a production-unit basis horticultural commodities are of much higher value than agronomic commodities. For instance, vastly more land is devoted to wheat than to cabbage, and the total value of wheat production is orders of magnitude greater than the total value of cabbage production. But wheat is a relatively low-value crop in comparison to cabbage; thus, a field of cabbage is much more valuable than a field of wheat.

Virtually all horticultural commodities represent high-value crops. As we discussed in Chapter 5, tolerating insect injury directly relates to crop value. With a high-value crop, even a minor amount of injury can cause an economic loss. Consequently, the horticultural crops are much more intolerant of insect injury than agronomic crops.

Besides crop values, horticultural and agronomic crops differ in some important aspects of production. Growing regions for some vegetable and fruit crops are restricted because of specific growing requirements. In addition, many of these areas are cropped all year, not seasonally as with agronomic crops. Consequently, life cycles for some horticultural pests may not be interrupted. Because horticultural commodities are high value, production practices tend to be more intense and exacting than those for most agronomic crops. Moreover, agronomic crop production is directed towards yield, but horticultural crop production is directed towards both yield and quality. An emphasis on crop quality has important implications for managing insect injury.

Tolerating Insect Injury

A central consideration in tolerating injury is the type of injury. Insects directly attacking yield-producing structures cause greater losses than those causing indirect injury, in which other plant parts are damaged. Defining injury as direct or indirect depends on how we define yield-producing structures. For example, an insect feeding on leaves of an apple tree is producing indirect injury (the yield-producing structures of an apple tree are apples). But an insect feeding on cabbage leaves produces direct injury (the yield-producing structure of a cabbage is the head which is composed of cabbage leaves). Many horticultural pests cause direct injury. Indeed, most important horticultural pests directly attack fruit or other yield-producing structures.

Differences among horticultural plants have implications for managing insect injury. One significant division among horticultural crops is that some are food plants, whereas others are ornamental plants. For the most part, injury to horticultural food plants can be understood by yield reduction. Insect injury causes a yield reduction and therefore an economic loss. However, injury to ornamental plants can not be described solely in terms of yield loss. Instead, most injury to ornamental plants is termed aesthetic injury.

Aesthetic Injury

Aesthetic injury is a reduction in the value of an animal, plant, structure, or area based upon judgments regarding appearance and desirability and resulting from insect effects or the presence of insects. Aesthetic injury may include such situations as the presence of insects in a dwelling, insect defoliation on leaves or flowers of landscape plantings, or insect webbing in plants. Unlike other types of injury in which some measurable loss in yield is determined, aesthetic injury involves opinions such as how insects and their activities reduce the beauty of a plant or the acceptability of a dwelling.

In many instances it is possible to associate an economic loss with some degree of aesthetic injury, however, this relationship between aesthetic injury and economic loss is extremely subjective (as is aesthetic injury itself). For example, paying a pest control operator to treat a house for cockroaches represents one type of economic cost arising from aesthetic injury. A less subjective situation occurs when aesthetic injury reduces the marketability of ornamental plants.

Frequently, people are intolerant of aesthetic injury. In an extreme case, the mere presence of an insect or insects on a plant is regarded as unacceptable. Consequently, substantial control efforts often are undertaken for pests causing aesthetic injury. These efforts at reducing aesthetic injury not only are directed at ornamental plants; horticultural food crops also must meet aesthetic criteria. Although many types of insect injury may not reduce yield or nutritional quality, some types of injury may reduce the acceptability (and therefore price) of fruits and vegetables. An orange with a discolored skin from mite feeding, although equivalent to an unblemished orange in taste and quality, would not command as high a price as the uninjured orange. Indeed, blemished fruit frequently is excluded from the fresh market and must be sold for juice or other processing (at a much lower price).

Pest Management for Horticultural Pests

Many features of horticulture and horticultural pests frustrate our use of pest management programs. The diversity in horticultural commodities and pests requires conducting research and developing management programs for many pest and production systems. Additionally, because consumers often require near perfect, unblemished produce, growers must consider aesthetic injury in managing insect pests. Reducing aesthetic injury to fruits and vegetables (sometimes called **cosmetic injury**) often involves preventative pesticide treatments so as to avoid any injury whatsoever. Consequently, managing aesthetic injury limits management options because very few or no insects can be tolerated. Similarly, high cash values for most horticultural crops and having many insects causing direct injury contribute to an intolerance for insect injury. Ultimately, this intolerance is reflected in very low economic injury levels (EIL's) or aesthetic injury levels for most horticultural pests. **Aesthetic injury levels** are levels of aesthetic injury (based on subjective assessments) that justify economic control.

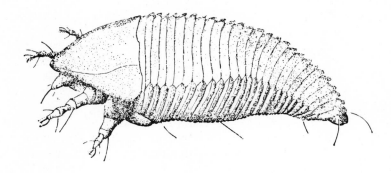

Figure 6.1   Citrus rust mite. These mites cause a russetting or browning of citrus fruit, a type of cosmetic injury (Courtesy USDA)

Management options for horticultural crops are similar to those for agronomic crops. For example, cultural techniques, biological control, and use of insecticides are as valuable for the management of horticultural pests as for agronomic pests. Alternatively, some control tactics, such as resistant cultivars, are more often used for disease management than for insect pest management with horticultural crops. Unfortunately, pest management may take a back seat to other production considerations for some horticultural crops. Consequently, some management options may be unavailable. Because most horticultural crops have low EIL's, proper scouting and sampling are crucial to avoid economic losses from a developing pest population.

HORTICULTURAL GROUPS

Vegetables and Mushrooms

Vegetables and mushrooms are an extremely important group of horticultural crops. Table 6.1 shows major vegetable crops in the US. Overall, the most important individual commodities are potatoes and tomatoes, which routinely have values of production (total values) in excess of a billion dollars. The importance of individual commodities varies among uses. The ten most important fresh market vegetables in the US are asparagus, broccoli, carrots, cauliflower, celery, lettuce, melons, onions, sweet corn, and tomatoes. The most important processed vegetable is tomato, and the most important frozen and canned vegetables are green peas, snap beans, and sweet corn.

Table 6.1　Major horticultural crops in the United States, 1980 & 1985: vegetables and mushrooms (fresh market and processed). (Data from USDA Agricultural Statistics, 1981 and USDA Agricultural Statistics, 1986, US Government Printing Office, Washington, D.C.; na = data not available.)

Crop	Harvested (thousands of acres) 1980	1985	Value (millions of $) 1980	1985
asparagus	82.9	na	82.1	na
beans, snap	350.9	222.2[1]	192.2	118.9[1]
broccoli	77.8	109.5	144.6	239.3
cabbage	96.5	na	175.2	na
carrot	72.4	90.1	161.4	206.4
cauliflower	43.3	61.2	95.8	169.1
celery	37.2	33.9	169.9	189.5
corn, sweet	552.6	625.3	263.6	368.1
cucumber	170.4	115.2[1]	183.5	123.6[1]
lettuce	242.6	225.7	565.2	674.7
muskmelon (cantaloupe)	86.3	na	161.1	na
onion	113.0	122.8	346.5	347.2
peas	321.7[1]	353.6[1]	101.4[1]	138.5[1]
peppers, green	55.5	na	123.7	na
potato	1182.5	1405.6	1720.4	1563.4
sweet potato	103.1	105.1	131.1	142.9
tomato	389.3	390.1	903.8	1195.5
watermelon	184.5	na	149.8	na
	Production (thousands of tons)			
mushrooms	235.0	297.8	368.5	493.6

[1]includes processed only.

Vegetables are among the highest cash value (value per production unit) crops. However, the cash value varies both among and within commodities. For example, potato has the lowest cash value of any vegetable, but the highest total value because potato production is greater than that of any other vegetable. A producer's price for a commodity also depends on its final use. Generally, fresh market vegetables bring higher prices than processed vegetables. Because crops with high cash values have relatively low EIL's, differences in cash value can significantly influence pest management.

The timing of crop production, specifically planting and harvesting dates, relates to both utilization and pest management. Frequently, vegetables are grown with the intent of harvesting at a specific time. Targeting harvest dates ensures that fresh market vegetables are available over a long period and that producers avoid sudden gluts in a commodity, which depress prices. Processed and preserved vegetables also may be planted to provide staggered harvest dates to allow processing plants to operate continuously over an extended period. Timing production for specific harvest dates also has important implications for pest management. Injury caused by some pests can alter the phenology of a crop, thereby changing harvest dates. Additionally, through extending the planting or harvesting period, a crop may be more exposed to injury by certain pests.

Pest management practices for home gardens and for commercial production are substantially different. Although timing and low EIL's are important constraints on pest management for commercial vegetable production, these factors are not as important for vegetable production in home gardens. Frequently, home gardeners tolerate more insect injury (higher EIL's) than do commercial producers. Additionally, intercropping (growing different crops in close association) and small production areas may reduce the potential for pest attack in gardens as compared to commercial fields. Moreover, in gardens many pests can be controlled through mechanical methods, such as picking insects off plants, whereas commercial production may depend more on insecticidal control.

Table 6.2   Major horticultural crops in the United States, 1980 & 1985: fruits (fresh market and processed) and nuts (in shell). (Data from USDA Agricultural Statistics, 1981 and USDA Agricultural Statistics, 1986, US Government Printing Office, Washington, D.C.; na = data not available.)

Crop	Production (thousands of acres)		Value (millions of $)	
	1980	1985	1980	1985
FRUIT				
apples	4409	3975	761.0	908.8
avocado	269	177	101.2	164.4
bushberries[1]	84	na	62.3	na
cherries	281	276	137.0	169.6
citrus:				
grapefruit	2986	2280	300.2	295.3
lemons	789	980	168.4	175.0
oranges	11832	6719	1304.2	1549.2
others[2]	877	515	102.2	130.4
total citrus	16484	10494	1875.0	2149.9
cranberries	135	171	89.5	191.9
grapes[3]	5595	5606	1341.0	960.6
nectarines	191	211	44.0	67.8
olives	109	96	40.1	51.1
peaches	1534	1074	367.2	308.5
pears	897	747	175.8	200.6
plums[4]	821	641	190.1	189.4
strawberries	351	509	288.8	450.8
NUTS				
almonds (shelled)	161.0	232.5	473.3	306.5
filberts	15.4	24.6	17.7	16.7
pecans	91.7	122.2	143.3	166.3
walnuts	197.0	219.0	184.4	161.4

[1]includes blackberries, blueberries, raspberries, and others.
[2]limes, tangerines, tangelos, and temples.
[3]includes raisins.
[4]includes prunes.

Fruits and Nuts

Fruits and nuts are the second major group of horticultural commodities (Table 6.2). Citrus, particularly oranges, is the most important US fruit crop, with a total value of production estimated at two billion dollars annually. Apples and grapes also are major fruit crops, each with annual values in excesses of one billion dollars. The value per production unit (cash value) of fruit crops is extremely high and is even more dependent on utilization than is the cash value of vegetables. In citrus, for example, fresh market fruit commands a much higher price than does fruit for juice. Because appearance standards for fresh market fruit are extremely high and because price differences between fresh market and processed fruit are substantial, minimizing aesthetic injury often is a primary consideration in fruit production.

Forecasting pest occurrence and sampling pest populations are particularly important pest management activities for fruit pests. Agronomic crops, for example, usually can tolerate some insect injury; therefore, precise estimates of the first occurrence of a pest may not be critical. In contrast, fruit crops have such extremely low EIL's that developing pest populations must be identified early to avoid economic losses. In many instances, preventative control measures are used to avoid all insect injury. Unfortunately, such practices can contribute to the development of insecticide resistance and to outbreaks of secondary pests. Although insect pest management programs may not eliminate the need for preventative treatments, they may reduce the number of unnecessary insecticide applications. Thus, accurate and timely monitoring of pest populations is vital in pest management for fruits.

Pest management for fruits and nuts involves both long-term and short-term considerations. Unlike most other crops we have considered, fruits and nuts are perennials, therefore the long-term health of the plant is a central concern. Indeed, production and pest management activities that promote good plant health will contribute to increased yields as well as to increased plant longevity. Although many insect pests directly attack fruit, many others affect trees through wood boring, twig and branch injury, or leaf injury. Pest management programs for fruit and nut trees must address both pests directly influencing fruit yield and pests that are more important in reducing tree health.

Turfgrass

Turfgrass, uniform grass stands kept to a low height and used for ornamental and recreational purposes, is an important feature of urban and rural landscapes. In 1983, the US had between 15 and 20 million acres of turfgrass. Perhaps somewhat surprisingly, expenses for turfgrass maintenance (including cutting, applying fertilizers, pesticides, and other activities) for the US are estimated at as much as $15 billion annually. Major locations for turfgrass are on golf courses, institutional lands, public lands, and residential properties, with approximately 70% of all expenditures for turfgrass going for residential properties. At least 40 major grass species are grown as turfgrass in the US including cool- and warm-season types.

Some insect pests of turfgrass produce aesthetic injury, however, others may kill sections or entire grass stands. Insect pests may inhabit one of three regions in the grass stand. Mite and aphid species occur on leaves and stems, which are the uppermost portions of the grass stand. Pest groups such as chinch bugs, cutworms, and webworms occur in the thatch. (**Thatch** is the accumulation of organic matter, dead grass, and debris below grass foliage and above the soil surface.) Other pests inhabit the soil and attack grass roots. These pests include mole crickets and white grubs (Fig. 6.2) that may attack grass in the spring or fall, as well as other pests such as chafers and Japanese beetles that are most damaging in the summer.

Pest management for turfgrass requires early recognition of symptoms and the proper identification of pests. Many common symptoms such as poor stand, dead turf, and yellowing can arise through many causes including disease, soil pH, and poor nutrition, as well as insects.

Consequently, proper identification of pests associated with a given problem is crucial to provide the correct response.

Many features of turfgrass production can modify the severity of insect problems. In particular, vigor of the grass stand influences the ability of turfgrass to withstand many types of injury. Additionally, some resistant cultivars are available. Nevertheless, insect control in turfgrass usually involves chemical methods and often includes preventative treatments (because of low tolerance for aesthetic injury to the grass). Management, including pest management, of turfgrass may differ with different uses. For example, athletic fields or golf courses are more intensively managed than are most home lawns. Management also may differ within a situation; for instance, on golf courses putting greens are more intensely managed than are fairways.

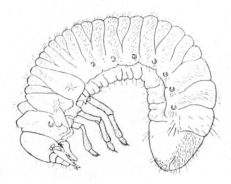

Figure 6.2   White grub, the larva of a scarab beetle. This insect attacks grass roots. (Courtesy Missouri Cooperative Extension Service)

Floricultural Crops

Floricultural crops include potted and cut flowers and bedding and foliage plants. Production and value of floricultural crops have increased over the last twenty years and this trend seems to be continuing. Table 6.3 summarizes the value of production for some major floricultural species in the U. S.; however, a great many more species of floricultural plants are grown than appear in Table 6.3.

Obviously, specific management practices differ among individual species, but many features of production are common across commodities. Floricultural plants are grown in fields, shade houses (or similar structures), and greenhouses, with most production in greenhouses. Some production is seasonal, but where possible, such as in California and Florida, plants are grown all year.

Production practices can be quite exacting and involved, including considerations of propagation, temperature, photoperiod, plant nutrition, and pests. Moreover, production of many species has high labor requirements. And consumer demand for floricultural plants is very strong. Consequently, floricultural plants are extremely high value crops, having possibly the highest cash value of any agricultural group.

Table 6.3   Major horticultural crops in the United States, 1980 & 1985: floricultural crops.   (Data from USDA Agricultural Statistics, 1981 and USDA Agricultural Statistics, 1986, US Government Printing Office, Washington, D.C.; na = data not available.)

Crop	Value (millions of $)	
	1980	1985
cut flowers:		
carnations	50.9	64.9
chrysanthemums	63.5	73.3
gladioli	21.8	25.4
roses	105.7	151.3
flowering potted plants:		
chrysanthemums	68.3	na
geraniums	41.8	na
lilies	19.2	na
poinsettias	66.1	na
bedding plants:		
flowering	124.8	na
vegetable	47.5	na
foliage plants	295.9	na

Individual floricultural crops frequently have associated specific insect pests; however, a number of pests attack many floricultural species. The most important greenhouse pests (aphids, mealybugs, spider mites, thrips and whiteflies) (Fig. 6.3) may attack almost any plant grown in greenhouses. Indeed, some of the pests listed as floricultural pests also are vegetable pests when vegetable crops are grown in greenhouses. These greenhouse pests have many common features that help account for their importance as pests. Most are small allowing easy entry through screens and small openings into the greenhouse. Most also have very short generation times and may even be parthenogenic; therefore significant pest populations can develop rapidly. Additionally, short generation times and frequent use of insecticides contributes to development of insecticide resistance in populations of many of these pest species.

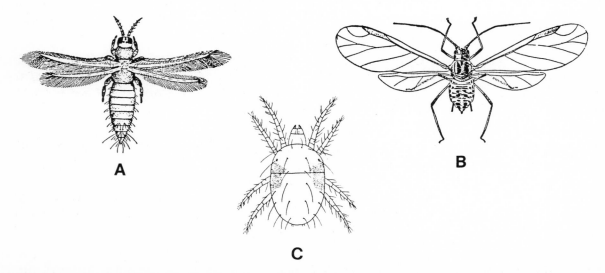

Figure 6.3   Some major greenhouse pests. A. Flower thrips. B. Green peach aphid. C. Twospotted spider mites. (A Courtesy Florida Agricultural Experiment Station; B Courtesy USDA; C Courtesy Illinois Cooperative Extension Service)

Insecticidal control is unquestionably the most important and prevalent tactic used against floricultural pests, but some other tactics are available. Besides insecticidal methods, barriers (either physical or chemical) prevent or reduce entry of pest species into greenhouses, and releases of biological control agents have been used against greenhouse pests, particularly in Europe. Plant resistance is an important tactic for many diseases but is less important for insects.

Because floricultural crops are essentially ornamental plants, aesthetic injury is an important pest management consideration. The need to minimize aesthetic injury coupled with the extremely high cash value of floricultural crops, constrains pest management options. In many production systems, timed pesticide sprays are used to control diseases and insects. However, pest management systems are available or under development to reduce the need for preventative treatments. One important feature of pest management in greenhouses is monitoring pest populations to time control measures, thereby reducing unnecessary insecticide applications.

OBJECTIVES

1) Understand how insect pest management for horticultural crops differs from pest management for other commodities. Know the limitations and potentials for pest management with high-value crops.

2) Understand features and differences in insect pests and pest management for types of horticultural crops including vegetables, mushrooms, fruits, nuts, turfgrass, and floricultural plants.

3) Learn the basic production characteristics and implications for pest management in the following commodities: potato, tomato, apple, citrus, grape, residential lawns, and cut flowers (any major species).

4) Know major life history features and be able to sight identify the following vegetable insect species or insect groups: bean leaf beetle, cabbage looper, Colorado potato beetle, leafhoppers, root maggots (cabbage, onion, and seedcorn maggots), spotted and striped cucumber beetle, tomato fruitworm, and twospotted spider mite.

5) Know major life history features and be able to sight identify the following fruit insect species or insect groups: apple maggot, cherry fruit fly, citrus thrips, codling moth, leafhoppers (grape leafhopper), Oriental fruit moth, plum curculio, and San Jose scale.

6) Know major life history features and be able to sight identify the following turfgrass insect species or insect groups: chafers, chinch bugs, Japanese beetle, mole crickets, and white grubs.

7) Know major life history features and be able to sight identify the following floricultural insect species or insect groups: aphids, greenhouse whitefly, mealybugs, mites, scales, and thrips.

8) Additional objectives (at instructor's discretion):

OPTIONAL DISPLAYS AND EXERCISES

1) Select two horticultural commodities not in objective 3 (your choice) and learn production practices and insect pests (including identification).

2) Determine recommended management practices for major vegetable and fruit pests grown in your area. Know how insect pests and management practices differ between commercial and home production.

3) Examine displayed groups of insects (which include pests and nonpests) and identify the insect pests.

4) Examine displayed examples or pictures of injury by horticultural insects.

5) Additional displays or exercises (at instructor's discretion):

TERMS

aesthetic injury	cosmetic injury	thatch	turfgrass
aesthetic injury level	horticulture		

MAJOR VEGETABLE INSECT PESTS

armyworms, especially:

beet armyworm	*Spodoptera exigua*	Lepidoptera: Noctuidae
yellowstriped armyworm	*Spodoptera ornithogalli*	Lepidoptera: Noctuidae
asparagus beetle	*Crioceris asparagi*	Coleoptera: Chrysomelidae
aster leafhopper	*Macrosteles fascifrons*	Homoptera: Cicadellidae
bean leaf beetle	*Cerotoma trifurcata*	Coleoptera: Chrysomelidae
cabbage aphid	*Brevicoryne brassicae*	Homoptera: Aphididae
cabbage looper	*Trichoplusia ni*	Lepidoptera: Noctuidae
cabbage maggot	*Delia radicum*	Diptera: Anthomyiidae
carrot rust fly	*Psila rosae*	Diptera: Psilidae
corn earworm	*Heliothis zea*	Lepidoptera: Noctuidae
= bollworm/tomato fruitworm		
European corn borer	*Ostrinia nubilalis*	Lepidoptera: Pyralidae
harlequin bug	*Murgantia histrionica*	Hemiptera: Pentatomidae
imported cabbageworm	*Artogeia rapae*	Lepidoptera: Pieridae
melon aphid	*Aphis gossypii*	Homoptera: Aphididae
= cotton aphid		
Mexican bean beetle	*Epilachna varivestis*	Coleoptera: Coccinellidae

onion maggot	*Delia antiqua*	Diptera: Anthomyiidae
onion thrips	*Thrips tabaci*	Thysanoptera: Thripidae
pea aphid	*Acyrthosiphon pisum*	Homoptera: Aphididae
pepper maggot	*Zonosemata electa*	Diptera: Tephritidae
pepper weevil	*Anthonomus eugenii*	Coleoptera: Curculionidae
phorid mushroom fly	*Megaselia halterata*	Diptera: Phoridae
pickleworm	*Diaphania nitidalis*	Lepidoptera: Pyralidae
potato aphid	*Macrosiphum euphorbiae*	Homoptera: Aphididae
potato flea beetle	*Epitrix cucumeris*	Coleoptera: Chrysomelidae
potato leafhopper	*Empoasca fabae*	Homoptera: Cicadellidae
sciarid mushroom fly	*Lycoriella mali*	Diptera: Sciaridae
seedcorn maggot	*Delia platura*	Diptera: Anthomyiidae
spotted cucumber beetle = southern corn rootworm	*Diabrotica undecimpunctata howardi*	 Coleoptera: Chrysomelidae
striped cucumber beetle	*Acalymma vittatum*	Coleoptera: Chrysomelidae
sweetpotato weevil	*Cylas formicarius elegantulus*	Coleoptera: Cuculionidae
tomato fruitworm = bollworm/corn earworm	*Heliothis zea*	Lepidoptera: Noctuidae
tomato hornworm	*Manduca quinquemaculata*	Lepidoptera: Sphingidae
twospotted spider mite	*Tetranychus urticae*	Acari: Tetranychidae

MAJOR FRUIT INSECT PESTS

apple maggot	*Rhagoletis pomonella*	Diptera: Tephritidae
avocado red mite	*Oligonychus yothersi*	Acari: Tetranychidae
blackheaded fireworm	*Rhopobota unipunctana*	Lepidoptera: Tortricidae
blueberry maggot	*Rhagoletis mendax*	Diptera: Tephritidae
California red scale	*Aonidiella aurantii*	Homoptera: Diaspididae
cherry fruit fly = cherry maggot	*Rhagoletis cingulata*	Diptera: Tephritidae
cherry fruitworm	*Grapholita packardi*	Lepidoptera: Tortricidae
citrus red mite	*Panonychus citri*	Acari: Tetranychidae
citrus rust mite	*Phyllocoptruta oleivora*	Acari: Eriophyidae
citrus thrips	*Scirtothrips citri*	Thysanoptera: Thripidae
codling moth	*Cydia pomonella*	Lepidoptera: Tortricidae
cranberry fruitworm	*Acrobasis vaccinii*	Lepidoptera: Pyralidae
European red mite	*Panonychus ulmi*	Acari: Tetranychidae
filbertworm	*Melissopus latiferreanus*	Lepidoptera: Tortricidae
fruittree leafroller	*Archips argyrospila*	Lepidoptera: Tortricidae
grape berry moth	*Endopiza viteana*	Lepidoptera: Tortricidae
grape leafhoppers	*Erythroneura* spp.	Homoptera: Cicadellidae
grape rootworm	*Fidia viticida*	Coleoptera: Chrysomelidae
hickory shuckworm	*Laspeyresia caryana*	Lepidoptera: Tortricidae
navel orangeworm	*Amyelois transitella*	Lepidoptera: Pyralidae
obliquebanded leafroller	*Choristoneura rosaceana*	Lepidoptera: Tortricidae
oriental fruit moth	*Grapholita molesta*	Lepidoptera: Tortricidae
Pacific spider mite	*Tetranychus pacificus*	Acari: Tetranychidae

peachtree borer	*Synanthedon exitiosa*	Lepidoptera: Sesiidae
peach twig borer	*Anarsia lineatella*	Lepidoptera: Gelechiidae
pear psylla	*Psylla pyricola*	Homoptera: Psyllidae
pecan nut casebearer	*Acrobasis nuxvorella*	Lepidoptera: Pyralidae
pecan weevil	*Curculio caryae*	Coleoptera: Curculionidae
plum curculio	*Conotrachelus nenuphar*	Coleoptera: Curculionidae
raspberry fruitworm	*Byturus* spp.	Coleoptera: Byturidae
raspberry sawfly	*Monophadnoides geniculatus*	Hymenoptera: Tenthredinidae
redbanded leafroller	*Argyrotaenia velutinana*	Lepidoptera: Tortricidae
redberry mite	*Acalitus essigi*	Acari: Eriophyidae
San Jose scale	*Quadraspidiotus perniciosus*	Homoptera: Diaspididae
strawberry root weevil	*Otiorhynchus ovatus*	Coleoptera: Curculionidae
strawberry rootworm	*Paria fragariae*	Coleoptera: Chrysomelidae
walnut husk fly	*Rhagoletis completa*	Diptera: Tephritidae

MAJOR TURFGRASS INSECT PESTS

Bermudagrass mite	*Eriophyes cynodoniensis*	Acari: Eriophyidae
black turfgrass ataenius	*Ataenius spretulus*	Coleoptera: Scarabaeidae
European chafer	*Rhizotrogus majalis*	Coleoptera: Scarabaeidae
hairy chinch bug	*Blissus leucopterus hirtus*	Hemiptera: Lygaeidae
Japanese beetle	*Popillia japonica*	Coleoptera: Scarabaeidae
masked chafer	*Cyclocephala* spp.	Coleoptera: Scarabaeidae
mole crickets	*Scapteriscus* spp.	Orthoptera: Gryllotalpidae
sod webworms	Crambinae spp.	Lepidoptera: Pyralidae
southern chinch bug	*Blissus insularis*	Hemiptera: Lygaeidae
white grubs	*Phyllophaga* spp.	Coleoptera: Scarabaeidae

MAJOR FLORICULTURAL INSECT PESTS

aphids, especially:		
green peach aphid	*Myzus persicae*	Homoptera: Aphididae
melon aphid	*Macrosiphum gossypii*	Homoptera: Aphididae
= cotton aphid		
potato aphid	*Macrosiphum euphorbiae*	Homoptera: Aphididae
rose aphid	*Macrosiphum rosae*	Homoptera: Aphididae
chrysanthemum gall midge	*Rhopalomyia chrysanthemi*	Diptera: Cecidomyiidae
fungus gnat	*Bradysia* spp.	Diptera: Sciaridae
greenhouse whitefly	*Trialeurodes vaporariorum*	Homoptera: Aleyrodidae
mites, especially:		
bulb mite	*Rhizoglyphus echinopus*	Acari: Acaridae
twospotted spider mite	*Tetranychus urticae*	Acari: Tetranychidae
mealybugs, especially:		
citrus mealybug	*Planococcus citri*	Homoptera: Pseudococcidae
longtailed mealybug	*Pseudococcus longispinus*	Homoptera: Pseudococcidae
solanum mealybug	*Phenacoccus solani*	Homoptera: Pseudococcidae

striped mealybug	*Ferrisia virgata*	Homoptera: Pseudococcidae
omnivorous leafroller	*Platynota stultana*	Lepidoptera: Tortricidae
rose midge	*Dasineura rhodophaga*	Diptera: Cecidomyiidae
scales, especially:		
fern scale	*Pinnaspis aspidistrae*	Homoptera: Diaspididae
Florida red scale	*Chrysomphalus aonidum*	Homoptera: Diaspididae
latania scale	*Hemiberlesia lataniae*	Homoptera: Diaspididae
thrips, especially:		
flower thrips	*Frankliniella tritici*	Thysanoptera: Thripidae
gladiolus thrips	*Thrips simplex*	Thysanoptera: Thripidae

VEGETABLE AND MUSHROOM INSECTS BY COMMODITY

Asparagus

asparagus beetle

Beans, Snap

bean leaf beetle, European corn borer, Mexican bean beetle, potato leafhopper, and seedcorn maggot

Carrot

aster leafhopper and carrot rust fly

Celery

no major insect pests, various occasional pests

Corn, Sweet

corn earworm and European corn borer

Cole Crops (including broccoli, cabbage, cauliflower, and others)

cabbage aphid, cabbage looper, cabbage maggot, harlequin bug, and imported cabbageworm

Lettuce

armyworms (especially beet armyworm and yellowstriped armyworm), aster leafhopper, cabbage looper, and corn earworm

Mushroom

phorid mushroom fly and sciarid mushroom fly

Muskmelon (Cantaloupe)

melon aphid, pickleworm, and spotted and striped cucumber beetles

Onion

onion maggot and onion thrips

Pea

pea aphid, seedcorn maggot, and twospotted spider mite

Pepper, Green

European corn borer, pepper maggot, and pepper weevil

Potato

Colorado potato beetle, potato aphid, potato flea beetle, and potato leafhopper

Sweet Potato

sweetpotato weevil

Tomato

tomato fruitworm and tomato hornworm

Watermelon

pickleworm and spotted and striped cucumber beetles

FRUIT INSECTS BY COMMODITY

Apple

apple maggot, codling moth, obliquebanded leafroller, Pacific mite, plum curculio, redbanded leafroller, and San Jose scale

Avocado

avocado red mite

Blackberry

redberry mite

Blueberry

blueberry maggot, cherry fruitworm, cranberry fruitworm, and plum curculio

Cherry

cherry fruit fly, European red mite, fruittree leafroller, and San Jose scale

Citrus

California red scale, citrus thrips, citrus red mite, and citrus rust mite

Cranberry

blackheaded fireworm and cranberry fruitworm

Grapes

grape berry moth, grape leafhoppers, and grape rootworm

Peaches and Nectarines

oriental fruit moth, Pacific spider mite, peach tree borer, peach twig borer, and San Jose scale

Pear

codling moth and pear psylla

Plum

Pacific mite, peach twig borer, and San Jose scale

Raspberry

raspberry fruitworm and raspberry sawfly

Strawberry

strawberry root weevil and strawberry rootworm

NUT INSECTS BY COMMODITY

Almond

navel orangeworm and peach twig borer

Filbert

filbertworm

Pecan

hickory shuckworm, pecan nut caseborer, and pecan weevil

Walnut

codling moth, navel orangeworm, and walnut husk fly

DISCUSSION AND STUDY QUESTIONS

1) How are insect pests and pest management likely to differ between tomatoes grown in the field and hydroponic tomatoes grown in a greenhouse?

2) Features of potato production make pest management for potato more similar to that of most agronomic crops than of most vegetable crops. Explain.

3) Mushrooms are grown in enclosed buildings with temperature and humidity control analogous to controls in greenhouses; however, greenhouse insect pests are not mushroom pests. Why aren't greenhouse pests a problem on mushrooms?

4) How would pest management for oranges grown for juice be different from pest management for oranges grown for fresh market?

5) Pest management for fruit, such as apples, is much more dependent on timely sampling than pest management of agronomic crops, such as alfalfa. Why?

6) Large yellow patches are a common symptom of many types of injury and stress in turfgrass, including insect injury. In general, how is an insect association with such injury determined?

7) Which principles of insect pest management are of great value in cut flower production? Which are not?

BIBLIOGRAPHY

Horticulture encompasses a tremendous diversity of plant species, production practices, and pests with a corresponding diversity in publications on horticulture. Provided below are some useful titles (note, however, that there are few comprehensive guides to horticultural insect pests).

American Phytopathological Society. Compendia of Plant Diseases (various editors and titles). St. Paul, MN
- a growing number of APS disease compendia address horticultural commodities. Compendia are currently available for: beet (includes insects), citrus, grape, ornamental foliage plants, pea, potato, rhododendron and azalea, rose, strawberry, sweet potato, and turfgrass.

Childers, N. F. 1983. Modern fruit science. Horticultural Publications, Gainsville, FL
- a good general treatment of fruit production (excluding citrus), but little information on diseases or insects.

Hanson, A. A., and F. V. Juska, eds. 1969. Turfgrass science. American Society of Agronomy, Madison, WI
- a multiauthored monograph on turfgrass with chapters on topics such as nutrition,

physiology, weeds, diseases, insects, athletic fields, putting greens, highway roadsides, and many others.

Holley, W. D., and R. Baker. 1963. Carnation Production. Wm. C. Brown Co., Dubuque, IA
- a thorough discussion of carnation production with some treatment of diseases and insects. Dated but still useful.

Joiner, J. N., ed. 1981. Foliage plant production. Prentice-Hall, Inc., Englewood Cliffs, NJ
- a comprehensive guide to all aspects of commercial foliage plant production. Includes information on individual plant species with notes on diseases and insects.

Mastelerz, J. W., ed. 1976. Bedding plants. Pennsylvania Flower Growers, University Park, PA
Mastelerz, J. W., and E. J. Holcomb, eds. 1982. Geraniums. Pennsylvania Flower Growers, University Park, PA
Mastelerz, J. W., and R. W. Langhans, eds. 1969. Roses. Pennsylvania Flower Growers, University Park, PA
- manuals on commercial flower production, including chapters on specific features of production including insects and diseases.

Shetlar, D. J., P. R. Heller, and P. D. Irish. 1983. Turfgrass insect and mite manual. The Pennsylvania Turfgrass Council, Inc., Bellefonte, PA
- a well-illustrated guide that includes information on insect life histories and descriptions of injury.

Shurtleff, M. C., T. W. Fermanian, and R. Randell. 1987. Controlling turfgrass pests. Prentice-Hall, Inc., Englewood Cliffs, NJ
- a comprehensive treatment of insect, disease, and weed problems in turfgrass. In addition to many pest descriptions, it includes information on good production and maintenance practices.

Tashiro, H. 1987. Turfgrass insect of the United States and Canada. Cornell University Press, Ithaca, NY
- an outstanding guide to turfgrass insects. A comprehensive text aided by superb, full-color photographs of insects and symptoms of injury.

University of California. IPM and Production Guides (various titles and authors). Davis, CA
- a model series of publications on pest management and production. All are beautifully illustrated and include comprehensive information on pests including insects. Focus of the manuals is on California, but information is applicable to other areas. Manuals treating horticultural subjects include: IPM guides to almonds, citrus, cole crops, potatoes, tomatoes, and walnuts; grape pest management; pear pest management; prune orchard management; and walnut orchard management.

Ware, G. W., and J. P. McCollum. 1980. Producing vegetable crops. 3rd ed. Interstate Printers and Publishers, Inc., Danville, IL
- an excellent summary of vegetable production. Many chapters are devoted to individual commodities and include some information on diseases and insects.

Westcott, C. 1973. The gardener's bug book, 4th ed. Doubleday and Co., Inc., Garden City, NY
- a compendium of vegetable insect pests, directed towards home gardeners. A valuable reference, but showing its age (particularly in reference to control measures).

Woodruff, J. G. 1979. Tree nuts. AVI Publishing Co., Inc., Westport, CT
- a thorough text on nut production in the U. S. In addition to general chapters, each major nut species is discussed in a separate chapter. Limited information is provided on pests.

Wuest, P. J., eds. 1982. Penn State handbook for commercial mushroom growers. Pennsylvania State University, State College, PA
- an excellent reference for commercial mushroom production. It includes chapters on virtually all aspects of production, including considerable information on pest management and is well illustrated with line drawings and color photographs.

7 FOREST, SHADE TREE AND RELATED PESTS

Pest management for woody plants varies among plants and uses. Pests and principles may differ for insects attacking ornamental plantings (like shrubs), Christmas trees, shade trees, or trees in forests. In some instances, such as Christmas tree production, pest management procedures are very similar to those for agronomic or vegetable crops. Likewise, managing pests of woody ornamentals and shade trees involves most of the same considerations as for other types of horticultural plants. However, **silviculture**, the production and maintenance of forests, requires pest management approaches distinct from other types of agriculture. Further, the great diversity in woody plant species grown, including conifers, broad-leaved evergreens, and deciduous plants, leads to a corresponding diversity in insect pest species. Because substantial information on biology, sampling, control, and other areas is required for management of each pest species, having many pest species complicates management of tree pests.

In Chapter 6, on horticultural plants, we discussed pests of fruit and nut trees. The major concern in fruit and nut production is in protecting the fruit, whereas pest management for Christmas trees, shade trees, and ornamentals must focus on aesthetic injury, and pest management of forest trees is directed toward wood quality and the long-term health of plants. Consequently, the importance of a specific type of insect injury will differ among these situations. Obviously, injury that kills or greatly weakens a tree will be important for any type of production. But less severe injuries can vary in importance. For example, minor defoliation of a fruit or forest tree may be of little immediate concern; however, defoliation of an ornamental shrub or shade tree may be regarded as serious injury. Similarly, fruit loss in shade or forest trees may not be considered especially serious, but direct injury in fruit trees is among the most damaging type of injury possible.

Insect Injury

Differences in the importance of injury highlights a recurring theme we have discussed in pest management: tolerating injury. Although injury causing plant death usually represents an absolute, other types and degrees of injury differ in importance, depending on how much injury can be tolerated. Understanding how different types of tree production influence how injury is tolerated will help explain differences in pest management for these different production systems. Because the type of insect injury directly relates to how much injury can be tolerated, it is important to recognize different injuries produced by insects.

Virtually all tree and shrub tissues can be attacked by insects. Insects often are classified as attacking leaves, twigs and buds, stems (trunks), roots, or fruit. More specific types of injury can occur within these broader categories. **Mines**, small tunnels of removed plant tissue caused by insect feeding, can occur in leaves, stems, and fruit. **Boring**, the production of large channels through plant tissues (twigs or stems), is similar to mining but is more severe and often associated with secondary infections by fungi. Another type of injury is the formation of galls on leaves, twigs, branches, or stems. **Galls** are swellings (actually, abnormal, undifferentiated plant cell growth) that develop in response to the presence of an insect or pathogen in plant tissues. **Girdling**, feeding completely around a twig, branch or small stem, breaks conductive tissue and may kill plants or plant parts. Insects may **defoliate** (completely consume leaf tissue), mine, or **skeletonize** (consume upper or lower surface of a leaf, leaving leaf veins intact) leaves. In addition to leaves, insects can destroy other plant parts including buds, fruit (seeds), and roots. Insects also can vector a number of important plant pathogens of trees.

SHADE TREE AND ORNAMENTAL PLANTS

Shade trees and woody ornamentals are important components of parks, streets, and homes. They beautify landscapes, provide habitat for wildlife, and bring nature into otherwise artificial, urban settings. Beyond the important aesthetic, and perhaps psychological, value of shade and ornamental plants, they can be of significant economic importance through increasing property values.

In 1976, 57 million trees grew along streets and roadways. Accounting for trees on private property increases the total to anywhere from 200 to 340 million trees. Based on annual expenditures for tree care, authorities estimate a 30 year old street tree is worth over $700 and a 50 year old tree over $2200. By using such estimates, US street trees are worth over $15 billion. Including trees on private property puts the value of all shade trees at well over $50 billion. (Statistics taken from Kielbaso and Kennedy, in Frankie and Koehler, 1983, Urban Entomology: Interdisciplinary Perspectives.)

Insect Pest Management

Pest management for shade tree and ornamental plants has many features in common with production of other types of horticultural plants. In particular, appearance of plants is a primary consideration for production and pest management. Modified economic injury levels have been developed for some ornamental plants which relate aesthetic criteria to economics. Additionally, efforts are underway to devise aesthetic injury levels to provide a more objective method for determining when to control insect pests. Nevertheless, pest management for shade trees and ornamentals often involves subjective assessments regarding how much injury or how many insects are unacceptable. Minor infestations of insects producing webbing or slight defoliation might not have any substantial long term effect on plant health but may be treated because the injury or webbing is unsightly.

Because shade trees and woody ornamentals are perennials, aesthetic injury is not of exclusive concern, and maintaining plant health is an equally important objective. Shade trees, particularly well-established trees, can be extremely valuable plantings. Similarly, ornamental plants usually reflect a significant investment in time and money. Consequently, while managing against aesthetic pests is problematic, pest management activities to maintain plant health usually are economical over the long term.

Many features of production of shade trees and ornamentals are pertinent to pest management. Proper production and management practices are fundamental to reducing insect and disease problems. Planting well-adapted cultivars, providing proper nutrition, proper pruning, and related practices all contribute to improved plant health. Besides production factors, conditions in urban environments, particularly air pollution, can degrade plant health and contribute to insect and disease problems.

Regularly monitoring plants to identify developing problems is necessary to avoid undue injury. Although chemical controls are commonly used for insect pests, some other approaches are available. In residential settings, pruning or mechanically removing insects may be sufficient for some pest problems. Additionally, in some species, resistant cultivars (particularly those with disease resistance, including resistance to insect-vectored diseases) are available. Often widespread planting of a single species or a few related species contributes to pest problems. The dominance of elm in urban plantings both contributed to the spread of Dutch elm disease and made effects of the disease seem more severe (because all shade trees in some areas were elm, loss of elm trees meant the loss of all shade trees). Therefore, diversity in plantings can reduce the spread of a pest, while avoiding the risk of catastrophic losses from a single pest.

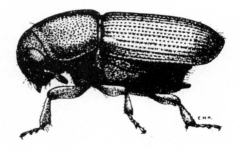

Figure 7.1   Smaller European elm bark beetle, a vector of Dutch elm disease. (Courtesy Oklahoma Cooperative Extension Service)

CHRISTMAS TREE PESTS

Christmas tree production is similar to commercial production of many horticultural plants in that appearance of the crop is a primary consideration for production and marketing. Specifically, Christmas trees are somewhat analogous to cut flowers, because both commodities are impermanent and sold based on aesthetic quality and consumer preferences. Consequently, practices to maintain and promote plant appearance are fundamental to Christmas tree production.

However, features of Christmas tree production and pest management probably have more in common with other types of tree production than with floricultural crop production. Christmas tree production requires several years and substantial acreages. A production period spanning many years allows time for trees to compensate for some types of injury. On the other hand, the long growing period lengthens the exposure of trees to insect and disease pests. And long growing periods can contribute to higher production costs, because a harvest occurs for a given acreage only once every few years, rather than one or more times a year.

Insect Injury

Insect injury to Christmas trees can take many forms, and many symptoms or signs may be associated with specific insect pests. Discolored foliage or loss of needles are two common symptoms associated with many diseases, adverse abiotic conditions, and insect injuries. Deformed tissues or abnormal growth, including galls, often denote a disease or insect infestation. Pitch secretions, wood shavings, or fine wood dust all indicate the presence of insects feeding in stems or shoots. Symptoms of injury or signs of insects are good indicators of pest problems, but such indications should be caught well before they become widespread. Consequently, timely tree inspections and a monitoring program for pests are necessary to avoid excessive and unnecessary injury.

Insect Pest Management

Good pest management for Christmas trees depends on good tree production practices. Planting sites should be well suited for trees and should not include frost-prone areas. Old trees or plant debris should be removed and weed control provided to improve seedling establishment. Species and variety selection are important early decisions for pest management, production, and marketing. Criteria include consumer preferences (for example, Scotch pines have long been a popular species), adaptation to soils and climate, and potential pest problems. When planting, both proper tree spacing and row orientation contribute to reduced disease problems later on, by increasing air flow around plants. Additionally, plant nutrition, water, and light all contribute to plant health, which in turn influences susceptibility of Christmas trees to many pests.

When developing pest problems are recognized, appropriate control methods should be undertaken. Insecticides are an important tactic for managing many Christmas tree pests, but other approaches are available and valuable. For instance, Scotch pine varieties resistant to many insects and diseases are available. Inspecting seedling transplants for insect pests and planting only pest-free material is another means to avoid some pest problems at the outset. Mechanical and cultural techniques, such as shearing and pruning (used to shape the trees) also reduce the need for insecticides. For example, delayed shearing can eliminate some infestations of European pine shoot moth. Similarly, pruning lower branches helps control pine root collar weevil and European pine shoot moth. Sanitation through collecting and destroying old branches and other plant debris reduces the occurrence of some insect pests. Biological control measures can be used against some pests, especially aphids and scales. And quarantines also have been successful in preventing the spread of a number of insect pests and diseases.

FOREST PESTS

Forests are a dominant feature of the North American continent. They are priceless ecological resources and are exceptionally important for recreation and commerce. Approximately one third of the United States is forested (Fig. 7.2). These forested lands are public and private, managed and unmanaged. In contrast to many other developed countries (particularly European countries), many American forests are undisturbed. In Great Britain, authorities believe no virgin forests (what the British call the wildwood) remain. Indeed, the last virgin woods were probably harvested in the 1100's and major deforestation began in Stone Age times, almost 10,000 years ago (Muir 1981, Riddles in the British Landscape). Thus, the unmanaged and virgin forests of the US are national treasures, as much so as managed, commercial forests.

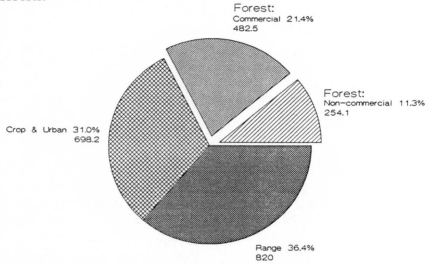

Figure 7.2 Uses of US land area, including Alaska and Hawaii (billions of acres).

Almost 66% (482.5 billion acres) of US forests are commercial. Of commercial forested land, 28% (135.7 billion acres) are public and 72% (346.8 billion acres) private (Fig. 7.3). US commercial timber production is summarized in Table 7.1. Beyond timber products listed, wood is valuable for other items, such as wood-derived chemicals (turpentine, paint, and varnish components). Notice that over 16% of annual production (growth) is lost through tree mortality, including effects of fire, disease, and insects.

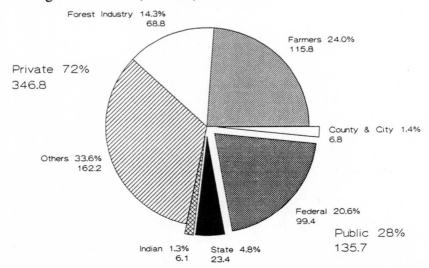

Figure 7.3 US commercial timberlands (billions of acres).

Table 7.1   US timber production (including Alaska and Hawaii). (Data from USDA-Forest Service, 1982, An analysis of the timber situation in the United States 1952-2030, Forest Resource Report No. 23.)

Commercial Trees[1]: (billions of board ft.)	Growth 74,621	Removal[2] 65,177	Mortality 12,188
Timber Products: (millions of board ft.)	Lumber 39.4	Plywood & Veneer 10.1	Panel Products[3] 8.7
	Woodpulp 45.1	Misc. Products[4] 20.6	Fuel 11.7

[1]Commercial trees defined as at least 12 ft. long, 9 in. diameter if softwood, 11 in. diameter if hardwood.
[2]Removal includes harvesting, cultural operations, and land clearing.
[3]Panel products include insulating board, particle board, fiberboard, and others.
[4]Misc. products include fence posts, logs, poles, ties, and others.

Insect Injury

Insect pests place limitations on timber production. Their most important effect is to kill or seriously weaken trees. Besides loss of timber, large volumes of dead trees increase fire hazards. Tree mortality also degrades watersheds and contributes to increased erosion. Besides directly causing tree mortality, insects can indirectly kill trees by vectoring tree pathogens. Often native tree species are especially susceptible to introduced pathogens. The classic example of a devastating disease vectored by insects is Dutch elm disease vectored by smaller European elm bark beetle and native elm bark beetle. In addition to killing trees, insect injury may reduce plant growth. Further, insects can degrade wood by discoloring it, causing twisted wood grain, and putting holes or tunnels in timber. Some types of injury, complete defoliation caused by gypsy moth, for instance, reduces the aesthetic appeal of a forest.

Insect Pest Management

Insect pest management for forests is different from pest management in most other systems. Certainly, some approaches for pest management in forests are the same as those for shade trees or Christmas trees, but practices for these commodities are directed toward plant appearance and longevity, not timber production. Moreover, Christmas tree production is focused on a much shorter time interval than forest production. In fact, length of production distinguishes forest production from all other agricultural production. Long production times (years or decades) greatly complicate economic assessments of insect injury. Consequently, economic injury levels, at least in the conventional sense, are not very useful for forest insects. Long production times allow greater opportunity for trees to compensate for insect injury, but plants also are exposed to pests for a longer period. Moreover, forests occupy such vast areas that pest management activities such as sampling or control actions are difficult undertakings.

Reductions in growth and reductions in wood quality caused by insects may warrant management actions. But in most instances, forests can tolerate significant insect injury without incurring undue losses. Thus, forests are extremely tolerant of insect injury, and most pest management programs are directed toward insects causing tree mortality or serious reductions in plant health. In rangelands a high tolerance to insect injury occurs because of very low crop values. In forests, a high tolerance to insect injury occurs because of long production times and vast acreages devoted to production. But both systems are analogous in that insect pest management programs tend to focus on situations in which many plants are killed. Consequently, insect outbreaks are a primary consideration in the management of forest insects.

Outbreaks

Outbreaks are high numbers of a single insect species in specific locations or regions causing severe injury or tree mortality. Usually outbreaks are limited in time or space, but insect populations can be widespread and persist for many years. Introductions and subsequent increases of new pests often spread more and persist longer than outbreaks of native pests. Although insect-vectored diseases often are as important as insect outbreaks, disease outbreaks may not require huge vector populations. Outbreaks are more likely in pure stands than in stands of mixed species and are more likely in planted stands than in natural stands. Additionally, factors degrading tree health, such as hail, drought, fire, or disease, may contribute to pest outbreaks because tree defenses against pests are weakened. Managing insect outbreaks may include **containment**, preventing spread of the outbreak to additional areas, or **suppression**, reducing insect numbers to prevent injury and stop the outbreak.

Insect Pest Management Practices

Surveys, estimates of pest presence and status in an area, are extremely important in managing forest pests, especially in managing pest outbreaks. Two approaches to surveys are detection and evaluation. **Detection** refers to discovering the presence of an insect species or a threatening pest population in an area. **Evaluation** refers to determining features of pest infestations in more detail, including estimates of pest numbers, pest population growth trends, limits of infestation, potential for economic losses, and other factors. Surveys are essential for identifying insect problems before huge losses occur and pest populations are so large or widespread that control is impossible. Because of the large land area devoted to forests, survey practices may involve the use of remote sensing, such as by airplane or satellites, as well as ground sampling.

Natural control through abiotic and biotic factors is important in restricting many insect populations. Biological control has been employed for various forest insects, and use of pathogens against some forest pests seems to be promising. Employing cultural and production practices to reduce pest problems is called **silvicultural control**. Some silvicultural control measures are stand mixture (by planting species mixtures), species selection (by planting species well adapted to an area and site), shortened rotations (to reduce overmature stands), and disposal of plant residue (to reduce habitat for pest species). Chemical controls can be useful against forest insects, but they can present a number of problems. Insecticides may not be sufficiently persistent to prevent populations from increasing again. Insecticides may not be sufficiently selective and may kill beneficials. And insecticides can pose threats through environmental contamination. However, some of these problems are circumvented by newer materials, such as microbial insecticides, which are selective and do not threaten the environment.

OBJECTIVES

1) Understand pest management for shade trees, woody ornamentals, and Christmas trees, including similarities and differences from other horticultural crops and forests.

2) Understand pest management for forests, including the importance of insect outbreaks.

3) Know major life history features and be able to sight identify the following tree insect species or groups: bagworm, bronze birch borer, eastern pine shoot borer, eastern tent caterpillar, fall webworm, gypsy moth, pales weevil, and scale insects.

4) Know major life history features and be able to sight identify the following forest insect species or groups: elm bark beetles (native elm bark beetle and smaller European elm bark beetle), gypsy moth, mountain pine beetle, southern pine beetle, spruce budworm, and white pine weevil.

5) Additional objectives (at instructor's discretion):

OPTIONAL DISPLAYS AND EXERCISES

1) Examine the displayed groups of insects (which include pests and nonpests) and identify the insect pests.

2) Examine displayed examples of injury by tree insects.

3) Determine major shade tree pests for your area and recommended pest management practices.

4) Additional displays or exercises (at instructor's discretion):

TERMS

boring
containment
defoliate
detection

evaluation
galls
girdling
mines

outbreak
silvicultural control
silviculture

skeletonize
suppression
surveys

MAJOR SHADE TREE AND ORNAMENTAL INSECTS

aphids	Aphidae spp.	Homoptera: Aphidae
azalea lace bug	*Stephanitis pyrioides*	Hemiptera: Tingidae
bagworm	*Thyridopteryx ephemeraeformis*	Lepidoptera: Psychidae
boxwood leafminer	*Monarthropalpus buxi*	Diptera: Cecidomyiidae
bronze birch borer	*Agrilus anxius*	Coleoptera: Buprestidae
carpenterworm	*Prionoxystus robiniae*	Lepidoptera: Cossidae
dogwood borer	*Synanthedon scitula*	Lepidoptera: Sesiidae
eastern tent caterpillar	*Malacosoma americanum*	Lepidoptera: Lasiocampidae
elm leaf beetle	*Pyrrhalta luteola*	Coleoptera: Chrysomelidae
fall cankerworm	*Alsophila pometaria*	Lepidoptera: Geometridae
fall webworm	*Hyphantria cunea*	Lepidoptera:Arctiidae
flatheaded appletree borer	*Chrysobothris femorata*	Coleoptera: Buprestidae
forest tent caterpillar	*Malacosoma disstria*	Lepidoptera: Lasiocampidae
gypsy moth	*Lymantria dispar*	Lepidoptera: Lymantriidae
lilac borer = ash borer	*Podosesia syringae*	Lepidoptera: Sesiidae
locust borer	*Megacyllene robiniae*	Coleoptera: Cerambycidae
native holly leafminer	*Phytomyza ilicicola*	Diptera: Agromyzidae
orangestriped oakworm	*Anisota senatoria*	Lepidoptera: Saturniidae
redhumped caterpillar	*Schizura concinna*	Lepidoptera: Notodontidae
scales, especially:		
cottony maple scale	*Pulvinaria innumerabilis*	Homoptera: Coccidae
oystershell scale	*Lepidosaphes ulmi*	Homoptera: Diaspididae
pine needle scale	*Chionaspis pinifoliae*	Homoptera: Diaspididae
spring cankerworm	*Paleacrita vernata*	Lepidoptera: Geometridae
twolined chestnut borer	*Agrilus bilineatus*	Coleoptera: Buprestidae
western tent caterpillar	*Malacosoma californicum*	Lepidoptera: Lasiocampidae
whitemarked tussock moth	*Orgyia leucostigma*	Lepidoptera: Lymantriidae
yellownecked caterpillar	*Datana ministra*	Lepidoptera: Notodontidae

CHRISTMAS TREE INSECTS

eastern pine shoot borer	*Eucosma gloriola*	Lepidoptera: Tortricidae
European pine sawfly	*Neodiprion sertifer*	Hymenoptera: Diprionidae
Nantucket pine tip moth	*Rhyacionia frustrana*	Lepidoptera: Tortricidae
pales weevil	*Hylobius pales*	Coleoptera: Curculionidae
pine root collar weevil	*Hylobius radicis*	Coleoptera: Curculionidae

| white pine weevil | *Pissodes strobi* | Coleoptera: Curculionidae |
| Zimmerman pine moth | *Dioryctria zimmermani* | Lepidoptera: Pyralidae |

MAJOR FOREST INSECTS

birch skeletonizer	*Bucculatrix canadensisella*	Lepidoptera: Lyonetiidae
bronze birch borer	*Agrilus anxius*	Coleoptera: Buprestidae
Douglas-fir tussock moth	*Orgyia pseudotsugata*	Lepidoptera: Lymantriidae
forest tent caterpillar	*Malacosoma disstria*	Lepidoptera: Lasiocampidae
gypsy moth	*Lymantria dispar*	Lepidoptera: Lymantriidae
hemlock looper	*Lambdina fiscellaria fiscellaria*	Lepidoptera: Geometridae
jack pine budworm	*Choristoneura pinus*	Lepidoptera: Tortricidae
larch casebearer	*Coleophora laricella*	Lepidoptera: Coleophoridae
larch sawfly	*Pristiphora erichsonii*	Hymenoptera: Tentredinidae
large aspen tortrix	*Choristoneura conflictana*	Lepidoptera: Tortricidae
mountain pine beetle	*Dendroctonus ponderosae*	Coleoptera: Scolytidae
native elm bark beetle	*Hylurgopinus opaculus*	Coleoptera: Scolytidae
pales weevil	*Hylobius pales*	Coleoptera: Curculionidae
redheaded pine sawfly	*Neodiprion lecontei*	Hymenoptera: Diprionidae
smaller European elm bark beetle		
	Scolytus multistriatus	Coleoptera: Scolytidae
southern pine beetle	*Dendroctonus frontalis*	Coleoptera: Scolytidae
spruce beetle	*Dendroctonus rufipennis*	Coleoptera: Scolytidae
spruce budworm	*Choristoneura fumiferana*	Lepidoptera: Tortricidae
Swaine jack pine sawfly	*Neodiprion swainei*	Hymenoptera: Diprionidae
western spruce budworm	*Choristoneura occidentalis*	Lepidoptera: Tortricidae
white pine weevil	*Pissodes strobi*	Coleoptera: Curculionidae

DISCUSSION AND STUDY QUESTIONS

1) What are some of the problems associated with applying insecticides in forests (both with insecticide and application procedures)?

2) What are some possible causes for insect outbreaks in forests?

3) What features of urban settings contribute to insect pest problems on shade trees and ornamentals?

4) How might acid rain influence insect pest problems in forests?

5) What criteria should homeowners use in deciding whether or not to control insects on ornamentals?

BIBLIOGRAPHY

Allen, D. C., and J. E. Coufal. 1984. Introduction to forest entomology. Syracuse University Press, Syracuse, NY
- a short introductory guide to entomology and forest insects.

Berryman, A. A. 1986. Forest insects. Plenum Press, New York, NY
- a textbook on forest entomology. Considerable emphasis on insect ecology as it pertains to forest entomology.

Benyus, J. M. 1983. Christmas tree pest manual. USDA-Forest Service, St. Paul, MN
- an excellent guide to Christmas tree pests. It includes information on pest management principles, production practices, and an extensive survey of pests with full color illustrations.

Drooz, A. T., ed. 1985. Insects of eastern forests. USDA-Forest Service, Misc. Pub. No. 1428
- a thorough, up to date guide to insect pests east of the 100th meridian. Substantial information on insect species with black and white photographs.

Furniss, R. L., and V. M. Carolin. 1977. Western forest insects. USDA-Forest Service. Misc. Pub. No. 1339
- similar to the guide edited by Drooz, but for insects west of the 100th meridian.

Johnson, W. T., and H. H. Lyon. 1988. Insects that feed on trees and shrubs. 2nd ed. Cornell University Press, Ithaca, NY
- a superb guide to insects of shade trees and ornaments. The book has color photographs and substantial information on individual insect species, as well as photographs of injury. A companion volume on diseases of trees and shrubs also is available.

Martineau, R. 1984. Insects harmful to forest trees. Multiscience Pub. Ltd. for Minister of Supply and Services, Canada
- another excellent guide to forest insects, with fine color plates and information on insect identification, life history, and injury.

Raupp, M. J., J. A. Davidson, C. S. Koehler, C. S. Sadof, and K. Reichelderfer. 1987. Decision-making considerations for aesthetic damage caused by pests. Bull. Entomol. Soc. Am. 34:27-32
- an excellent article that explores the dilemma of when to control aesthetic pests. The authors describe a technique for addressing aesthetic pests in the context of an economic injury level.

8 MEDICAL AND VETERINARY PESTS

This chapter focuses on insect pests of humans (**medical pests**) and domesticated animals (**veterinary pests**). Although agronomic and horticultural insects routinely disturb food production, their effects are relatively minor compared to the economic disruption, suffering, and death caused by medical pests. Thus, the medical pests are an extraordinarily important group of insect pests. Insect pests of man have had a tremendous impact on human history and this influence continues. Diseases carried by insects dominate many parts of the world, particularly central Africa. In fact, some authorities have suggested that although humans evolved in Africa, civilization instead developed in the Middle East, Europe, and Asia because humans were able to escape disease. Nevertheless, even where continual disease could be avoided, periodic episodes could not. Epidemics of arthropod-born disease undoubtedly contributed to the fall of the Roman Empire. And transmission of plague by fleas led to the pandemic of the Middle Ages which killed a fourth to a third of the population of Europe. But probably the most significant insect transmitted disease is malaria, which continues to be one of, if not the most, important threat to human health worldwide.

Just as human welfare is at risk from insects, so is the health of wild and domesticated animals. Although some insect species attacking humans also attack other animals, many veterinary and medical pests are host specific. However, injuries produced by these pests have many common features regardless of whether a human or other animal host is being injured. In the following sections we will consider these injuries. How medical and veterinary pests affect their hosts is tremendously important in managing those insects. In addition to focusing on injury, we need to recognize how medical and veterinary pests differ. Consequently, we also will examine how pest management is different for these pests, as well as how it differs from the management of plant pests.

DIRECT EFFECTS

Insects may directly injure an animal host in many ways. Some types of injury may be caused by insect feeding, however, other insect activities may also be damaging. These effects frequently have recognizable economic consequences. However, direct effects, whether on humans, livestock, or other animals, also have less quantifiable results, including pain and suffering. There are six major categories of direct effects from insects:

(1) **annoyance** (and blood loss) - annoyance comes from disruptive activities of insects, such as flying around or landing on the head, and from feeding, possibly causing a blood loss (called exsanguination). Insects usually do not remove sufficient blood to cause a medical problem, although anemia and significant blood loss caused by insects have been documented with livestock. Nevertheless, annoyance is not a trivial effect of insects. Human activities frequently are disrupted by insects, and in some instances, such as when recreational facilities cannot be used because of insects, annoyance can cause substantial economic losses. With livestock, annoyance is of even greater importance. Continuous irritation from insects may reduce weight gain in cattle, may disrupt milk production, and may contribute to increased

susceptibility to other stresses. Many of the important livestock pests cause annoyance.

(2) **dermatosis** (and dermatitis) - dermatosis is a disease of the skin, dermatitis an inflammation of the skin. Both dermatosis and dermatitis can be caused by arthropod activities. Many mite species, such as scabies mites and chiggers, produce acute skin irritations. **Human scabies**, a skin disease caused by infestations of the itch mite (*Sarcoptes scabiei*) (Fig. 8.1), is an important public health problem and periodic outbreaks are common. In livestock, **mange**, any persistent skin inflammation (often with accompanying hair loss) caused by mites, can seriously weaken animals. Serious, debilitating mange conditions in livestock are called **scabies**.

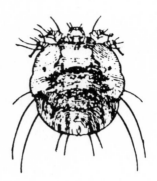

Figure 8.1   Itch or scabies mite. (Courtesy US Public Health Service)

(3) **myiasis** - is the invasion and feeding on living tissues of humans or animals by dipterous larvae. Fortunately, myiasis is a rare condition in humans, but it commonly occurs in livestock (Fig. 8.2). Besides the detrimental effects of myiasis itself, many additional complications can arise from myiasis, such as secondary microbial infections, secondary infestations by other insects, and debilitation. Myiasis can be fatal.

Figure 8.2   Example of myiasis, a cattle grub larva living and feeding under the skin of a cow. (Courtesy USDA)

(4) **envenomation** - is the introduction of a poison into the body of humans and animals. Few arthropods have sufficiently toxic poisons to kill humans outright. However, humans and other animals do die from arthropod venoms, and envenomation can cause a variety of nonlethal effects. Five mechanisms are associated with envenomation: biting, as occurs with spiders; stinging, as occurs with scorpions and some Hymenoptera (such as ants, bees, and wasps); contact - passively or inadvertently touching a poisonous feature, such as

urticating hairs (hairs that produce wheals and itching) which are found on many Lepidoptera larvae and some spiders (such as tarantulas); active projection - contacting poisons that are secreted or expelled such as **vesicating fluids** (acid or alkaline liquids causing skin irritation or blistering), that occur in blister beetles; and **ingestion** - accidentally eating poisonous insects (e.g., horses can be killed by ingesting hay containing dead blister beetles).

(5) **allergic reaction** (anaphylaxis) - a hypersensitive response to insect proteins. All of the mechanisms associated with envenomation can also cause exposure to allergens. In fact, human deaths from bee and wasp stings usually are associated with hypersensitive reaction rather than direct effect of a toxin. Additionally, allergies to insect proteins may be expressed in other ways. For example, one study of individuals allergic to chocolate discovered that 37% of people tested actually were not allergic to pure chocolate but were allergic to cockroaches (cockroach parts are a common contaminant of cocoa - something to think about the next time you eat a candy bar).

(6) **entomophobia** - an irrational fear of insects. This may range from unwarranted fears of innocuous insects to sensory hallucinations. One extreme form of entomophobia is delusory parasitosis, in which individuals become convinced they are infested with insects when no infestation exists. Delusory parasitosis may even be manifested by physical symptoms such as skin irritations and welts. Entomophobia may cause undue alarm and anxiety, lead to unwarranted use of insecticides, and, in severe cases, requires professional treatment. To a certain extent, the common dislike and repulsion most people have towards insects also is an unwarranted fear and has the unfortunate consequence of increasing intolerance to insects and insect injury which leads to increased, and even unnecessary, use of insecticides and other management tactics.

INDIRECT EFFECTS

The primary indirect effect of medical and veterinary insects is disease transmission. Indeed, disease transmission is more important than any other effect produced by medical and veterinary pests. Understanding the relationship of arthropods to disease requires a consideration of many concepts and much terminology.

Organisms that produce disease are called **pathogens**, and **disease** itself is a stress condition produced by the effects of a pathogen on a susceptible host. Arthropods capable of transmitting pathogens are called **vectors**. Some diseases may depend on only a single host and a vector; however, other diseases may include multiple host species, and even multiple vectors. In any of these instances, an organism that maintains the infective agent (the pathogen source) when active transmission does not occur is termed a **reservoir**. For example, the reservoir for malaria is human populations, with transmission occurring when a mosquito feeds on a infected individual and later feeds on an uninfected individual. With plague, the most common reservoirs are rats and other rodents, with transmission occurring when fleas feed on rats or rodents and then feed on humans. Often, the infection in the reservoir species is less severe than in the primary host, however, this is not always the case (e.g., plague is as deadly to rats as it is to humans).

The study of the nature of disease, especially how a pathogen produces disease by altering host physiology, is the province of **pathology**. Another fundamental consideration in characterizing any disease is **epidemiology**, the study of the incidence, distribution, and determinants of disease in a population. In considering epidemiology we can recognize different levels and distribution of disease: **endemic** refers to disease being native to a region or population,

epidemic refers to disease outbreaks affecting a high proportion of a population, and **pandemic** refers to disease outbreaks affecting a wide geographical area and a high proportion of a population or populations.

Epidemiology is particularly important in describing the involvement of arthropods in disease transmission. In particular, understanding host/pathogen, vector/host, and vector/pathogen relationships is central to most epidemiological questions.

Host/Pathogen Relationships

Fundamentally, disease is a manifestation of interactions between host and pathogen. An array of environmental and physiological factors may influence these interactions. Additionally, qualities of the host and pathogen influence disease development. **Resistance** refers to a host's ability to prevent infection and disease; **virulence** refers to a pathogen's ability to produce disease. These terms apply equally to plant pathogens as to animal pathogens; however, practical implications of resistance are different for plants and animals. Whereas genetic resistance to disease is an important component to managing plant disease, selecting resistant genotypes has more limited applicability with livestock and is impossible for humans. However, conferring resistance through the use of vaccines is possible for humans and other animals and is a primary mechanism of disease management.

Host/pathogen relationships also are disrupted with various therapeutic agents. For example, plague infections can be treated with tetracycline and related antibiotics. Unfortunately, just as we may observe ecological backlash by insect populations to insecticides, so do many pathogen populations develop resistance to various drugs. Strains of the plasmodium causing malaria, for example, are resistant to antimalarial drugs such as chloroquine.

Vector/Host Relationships

Many aspects of insect behavior and life history are important in disease transmission, especially those relating to relationships between vectors and hosts. Generally, the closer the association between vector and host, the greater the suitability of the vector to transmit disease. Different degrees of association are possible. Species that live on or in a different species are called **parasites**; external parasites are called **ectoparasites**, and internal parasites are called **endoparasites**. If a parasite can only live on a given host species the relationship is called **obligate**, e.g., head lice are obligate ectoparasites of man. Alternatively, if a parasite does not live exclusively on a given host species, then the relationship is said to **facultative**, e.g., cat fleas are facultative parasites of humans. Additionally, some parasites may be continuous on a host (like lice) but others may be temporary (like fleas).

The association of a vector species to humans is crucial to the importance of medical pests. Animals living in close association with people are said to be **synanthropic**. Species that "like" (usually feed on) humans are called **anthropophilic**. Behavioral relationships to man can greatly influence the medical importance of a vector species. Both ticks and mosquitoes are facultative, blood-sucking parasites that vector a tremendous array of human pathogens. However, ticks are only incidentally associated with humans, whereas many mosquito species are anthropophilic and routinely feed on humans. These differences help explain why the incidence of tick-borne disease is relatively trivial compared to that of mosquito-borne disease.

Vector/Pathogen Relationships

The ability of a pathogen to survive and remain infective in or on a vector species is a critical factor in disease transmission. Two mechanisms of transmission are possible. **Mechanical transmission** is the transfer of pathogen from an infectious source to a susceptible host by a vector, without any reproduction or developmental changes in the pathogen. Generally, mechanical transmission is an inefficient mechanism for disease transmission. Many insects carry disease producing pathogens on their body parts, but relatively few are known to be associated with disease outbreaks. Table 8.1 summarizes important mechanically-transmitted diseases with arthropod vectors.

Table 8.1   Some major pathogens mechanically transmitted by arthropods.

Pathogen	Disease	Vector	Victim
Bacteria			
Anaplasma marginale	anaplasmosis	Tabanidae	cattle
Bacillus anthracis	anthrax	Tabanidae	humans, various animals
Francisella tularensis	tularemia	Tabanidae	humans
Salmonella spp.	salmonellosis	house fly, cockroaches, & *Hippelates* spp. (eye gnats)	humans
Shigella spp.	shigellosis	house fly	humans
Treponema pertenue	yaws	house fly, *Hippelates* spp. (eye gnats)	humans
Vibrio comma	cholera	house fly	humans
Protozoa			
Entamoeba histolytica	amebic dysentary	cockroaches, house fly	humans
Toxoplasma gondii	toxoplasmosis	cockroaches	humans, cats
Trypanosoma spp.	trypanosomiasis (various specific diseases)	Tabanidae	humans
Virus			
fowl pox virus	fowl pox	mosquitoes	birds
myxomatosis virus	myxomatosis	mosquitoes	rabbits

The other transmission mechanism is **biological transmission**, in which the pathogen either reproduces, undergoes developmental changes, or both in the vector. Biological transmission is the most effective and significant mechanism for disease transmission by arthropods. Table 8.2 presents important biologically-transmitted diseases arranged by arthropod vectors.

Frequently the relationships between vectors, pathogens, and hosts are complex, and the challenge in epidemiology is to resolve these complexities. For example, we mentioned the relationship between rats, fleas, and humans in plague transmission, but the plague pandemic of the 1300's resulted from more than transmission of pathogen from rat by flea to humans. In the Middle Ages rats were the reservoir for plague but also were susceptible to the pathogen. As rats were killed by plague, rat fleas left their hosts and looked for alternative hosts, usually humans. The plague pathogen (a bacterium, *Yersinia pestis*) was highly virulent and the European populace highly susceptible. Although these points account for how plague was introduced to human populations, they probably cannot account for the extremely rapid spread of plague through Europe. Probably many of the human plague victims developed a form of the disease called pneumonic plague, in which the infection is centered in the lungs and is easily transmitted by coughing. Thus, it is likely that once the disease was established, humans were themselves the most important vectors of plague.

Table 8.2 Some major pathogens biologically transmitted by arthropods.

Vector	Pathogen	Disease	Distribution[1]	Victim
ACARI (ticks & mites)				
	Rickettsia			
Macronyssidae:				
Liponyssoides sanguineus	*Rickettsia akari*	rickettsial pox	world	humans
Trombiculidae:				
Leptotrombidium spp.	*R. tsutsugamushi*	scrub typhus	SE Asia	humans
Ixodoidea (ticks):				
	Bacteria			
Argas persicus (& other spp.)	*Borrelia anserina*	avian spirochetosis	Af, Au, ME, SA	birds
Ixodes dammini	*B. burgdorferi*	Lyme disease	US	humans
Ornithodoros spp.	*B. recurrentis*	relapsing fever	Af, As, Eu, NA, & SA	humans
various spp.	*Francisella tularensis*	tularemia	world	humans
	Protozoa			
Boophilus annulatus (& other spp.)	*Babesia bigemina*	Texas cattle fever	Af, CA, Eu, SA, SEA (formerly US)	cattle
Rhippicephalus appendiculatus (& other spp.)	*Theileria parva*	East Coast fever	Af	cattle
	Rickettsia			
various, especially *Dermacentor* spp.	*Rickettsia rickettsii*	Rocky Mt. spotted fever	NA, SA	humans
	Virus			
Dermacentor andersoni (& other species)	CTF virus	Colorado tick fever	US, Can	humans
Hyalomma marginatum marginatum	CCHF virus	Crimean-Congo hemorrhagic fever	Af, As, Eu	humans
Ixodes persulcatus (& other spp.)	RSSE virus	Russian spring-summer encephalitis	USSR	humans
I. ricinus	LI virus	louping ill	Eu	sheep
I. ricinus (& other species)	TBE virus	tick-borne encephalitis	Eu (USSR)	humans
Ornithodoros porcinus	ASF virus	African swine fever	Af, Eu	swine
INSECTA (insects)				
ANOPLURA (lice)				
	Bacteria			
Pediculus humanus humanus	*Borrelia recurrentis*	epidemic relapsing fever	Af (formerly Eu)	humans
	Rickettsia			
P. humanus humanus	*Rickettsia prowazekii*	louse-borne typhus	Af	humans
P. humanus humanus	*Rochalimaea quintana*	trench fever	Af, As, CA, Eu, & SA	humans
DIPTERA				
	Virus			
Ceratopogonidae (punkies):				
Culicoides spp.	bluetongue virus	bluetongue	Af, A, US	sheep
Culicoides spp.	BEF virus	bovine ephemeral fever	Af, Au	cattle
Culicidae (mosquitoes):				
	Nematodes (filarial)			
Aedes, Anopheles, & *Mansonia* spp.	*Brugia malayi*	brugian filariasis	SEA	humans
Culex pipiens, Aedes & *Anopheles* spp.	*Wuchereria brancrofti*	Bancroftian filariasis	world (tropics)	humans
various spp.	*Dirofilaria immitis*	dog heartworm	world	dogs

[1]Af=Africa, As=Asia, Au=Australia, CA=Central America, Can=Canada, Eu=Europe, ME=Middle East, NA=North America, SA=South America, SEA=Southeast Asia, US=United States, and USSR=Soviet Union.

Table 8.2 Some major pathogens biologically transmitted by arthropods, continued.

Vector	Pathogen	Disease	Distribution[1]	Victim
Culicidae (mosquitoes):	**Protozoa**			
Anopheles spp.	*Plasmodium falciporum, P. malariae, P. ovale, & P. vivax*	malaria	world (tropics)	humans
	Virus			
Aedes spp., esp. *A. egypti*	DEN virus	dengue	Carribean,SEA	humans
Aedes spp., esp. *A. egypti*	YF virus	yellow fever	Af, CA, SA	humans
Aedes spp., esp. *A. triseriatus*	LAC virus	La Crosse encephalitis	NA	humans
Culex spp.	SLE virus	St. Louis encephalitis	NA, SA	humans, horses
Culex spp. (esp. *C. tritaeniorhychus*)	JBE virus	Japanese encephalitis	eastern As	humans
Culex & Culiseta spp.	WEE virus	western equine encephalitis	Eu, NA, SA	humans, horses
various spp.	EEE virus	eastern equine encephalitis	NA, SA	humans, horses
various spp.	VEE virus	Venezuelan equine encephalitis	NA, SA	humans, horses
various spp.	RVF virus	Rift Valley fever	Af	humans, livestock
Muscidae:	**Protozoa**			
Glossina spp.	*Trypanosoma brucei gambiense (=T. gambiense & T. rhodesiense)*	sleeping sickness	tropical Af	humans
Glossina spp.	*T. brucei brucei, T. congoleuse, & T. vivax*	nagana	tropical Af	cattle
Psychodidae (Phlebotominae):	**Bacteria**			
Lutzomyia verrucanum	*Bartonella bacilliformis*	Carrion's disease	SA	humans
Phlebotomus & Lutzomyia spp.	*Leishmania* spp.	leshmaniasis (including Kala-Azar, oriental sore)	world (tropics)	humans
	Virus			
Phelbotomus papatasi (& other spp.)	sand fly fever virus	sand fly fever	Af, As, Eu	humans
Simuliidae(black flies):	**Nematodes** (filarial)			
Simulium spp.	*Onchocerca volvulus*	onchocerciasis	Af, tropical SA	humans
Tabanidae:				
Chrysops spp.	*Loa loa*	loiasis	Af	humans
HEMIPTERA	**Protozoa**			
Triatominae spp.	*Trypanosoma cruzi*	Chagas' disease	SA	humans
SIPHONAPTERA	**Bacteria**			
Xenopsylla spp., primarily *X. cheopis*	*Yersinia pestis*	plague	world	humans
	Rickettsia			
various spp.	*Rickettsia typhi*	murine (flea-borne) typhus	CA, Eu, NA	humans
	Virus			
Spilopsyllus cuniculi	myxomatosis virus	myxomatosis	Au, Eu	rabbits

[1]Af=Africa, As=Asia, Au=Australia, CA=Central America, Can=Canada, Eu=Europe, ME=Middle East, NA=North America, SA=South America, SEA=Southeast Asia, US=United States, and USSR=Soviet Union.

This example illustrates many aspects of epidemiology. Although rats were a preferred host of the rat flea, as the plague bacillus killed the rats the fleas were forced to seek alternative hosts. Because rats are synanthropic, the most available alternate hosts were humans. Additionally, the relationship of the pathogen to the vector contributed to the effectiveness of rat fleas in transmitting plague. Once fleas ingested the pathogen, the bacillus multiplied in the gut. Eventually, bacillus would almost block the gut, and the infected fleas began to starve. In feeding, starving fleas would suck so forcefully that when the sucking muscles relaxed, recoil in the esophagus shot bacillus-laden blood back through the feeding tube into the host's blood stream. Thus, the combined vector, pathogen, and host relationships associated with plague were all conducive to plague epidemics and pandemics. Indeed, these conditions remained sufficiently favorable for plague epidemics to occur periodically through to the 1900's. Plague still occurs, but now the disease is easily treated with antibiotics if caught early on.

PEST MANAGEMENT FOR MEDICAL AND VETERINARY PESTS

Although pest management for medical and veterinary pests shares common features with pest management of plant pests, there are more differences than similarities. The basic distinction between the two systems is that in most instances fewer medical or veterinary pests can be tolerated than plant pests. In other words, the economic injury levels (EIL's) for medical and veterinary pests are lower than for plant pests. Additionally, more management techniques are available for plant pests than for medical and veterinary pests. Moreover, the use of a given technique may be more limited with medical and veterinary pests than with plant pests. For example, although host plants can be treated with insecticides to control pests, usually humans cannot be so treated (although insecticides are used for louse control on humans). And because EIL's are so low for medical and veterinary pests, the value of biological control is diminished, because biological control agents may not suppress pest populations sufficiently.

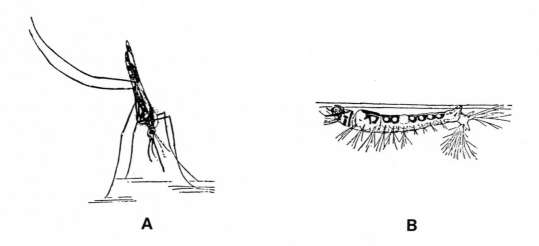

A   **B**

Figure 8.3 *Anopheles* mosquito, the vector of human malaria. A. Adult. B. Larva. (Courtesy USDA)

Management may differ for pests producing direct effects versus those with indirect effects (disease transmission). For pests with direct effects some level of pests may be tolerable, even if only a small number. Two approaches available for managing these pests are to avoid pests, through use of physical barriers or chemical repellents, and to reduce pest numbers. Although barriers are useful for medical insects, they are less useful for veterinary pests. Reducing pest populations also presents problems. Attacking active, flying insects may require treatment of large areas or developing methods to expose parasites to insecticide while on the host. Often immature pest stages may provide better opportunities for pest population reduction, because immatures are less mobile and may be less widely dispersed. Additionally, sanitation may be a useful method for some species, depending on manure or decaying material as larval habitats. 'Environmental manipulations, such as draining larval mosquito habitats, have been valuable in controlling some medical pests.

For indirect injury the situation is analogous to that of insects vectoring plant diseases - few or no pests can be tolerated. The objective in managing arthropod vectors is to prevent disease transmission and development. One approach to interrupting disease transmission is to change the vector/pathogen relationship. For example, releasing mosquito strains incapable of transmitting malaria has been suggested as one method to reduce malaria incidence. More commonly, disease transmission can be disturbed in one of two ways: by disrupting the activities of the vector or by disrupting the activities of the pathogen. Approaches for reducing vector activities are the same as discussed for reducing direct effects, i.e., barriers or reducing vector populations. These approaches pose some particular problems in that barrier techniques must completely exclude vectors and pest populations must be reduced to extremely low levels. If possible, disrupting the activities of the pathogen is a more favorable approach. Vaccines and therapeutic agents are the best techniques, but these are not available for many arthropod-borne diseases. Even when vaccines are available epidemics are still possible, because money, facilities, and staff to provide vaccine and vaccinations are not available in many parts of the world.

Medical Pest Management

Because the fundamental threat posed by medical pests is disease transmission, management of many of these arthropods presents virtually intractable problems. In temperate regions, some diseases have been eliminated or greatly diminished through reducing vector populations. For example, malaria no longer occurs in most of Europe and North America. But malaria and many other arthropod-borne diseases persist in the tropics. Although attempts have been made to eradicate insect vectors, these efforts have not been very successful. In fact, by subjecting pest populations to heavy, continual insecticide pressure, insecticide resistance has developed in a number of medically-important species. More balanced, rational management tactics may circumvent resistance problems, but the dilemma we face is that unacceptably high levels of disease may persist even after our best efforts at vector management. Potentially, focusing efforts on the pathogen through vaccine development or similar directions may offer more promise in controlling infectious disease. However, we are far from having a vaccine for many important infectious diseases.

Livestock Pest Management

Some of the concerns related to medical pests also apply to livestock pests. However, livestock pest management in the US more commonly relates to managing annoying insects. Table 8.3 indicates important livestock and livestock products in the US and their value. Arthropod pests cause significant losses in livestock production. Consequently, managing livestock pests is an important objective for any livestock producer.

Table 8.3   Livestock and livestock products in the United States, 1980 & 1985. (Data from USDA Agricultural Statistics, 1986, US Government Printing Office, Washington, D.C.)

Group	Millions of Units		Value (millions of $)	
	1980	1985	1980	1985
all cattle	111.2	109.7	55,844	44,139
all chickens	4,355.1	4,847.0	5,040	6,380
all hogs	64.5	52.3	4,815	3,639
all sheep	12.7	10.4	993	638
all turkeys	165.2	185.3	1,272	1,819
eggs	69,686.0	68,407.0	3,268	3,253
milk	128,406.0	143,667.0 (lbs.)	16,876	18,453
wool and mohair	115.3	101.3 (lbs.)	124	102

Although production practices and pest species differ among livestock species, some generalizations do apply. Two broad types of production systems are possible: confinement or range and pasture. In some situations a given system may include features of both production schemes, for example, pasture or open range in the summer, with confinement in the winter. In confinement systems livestock densities are high; therefore rapid movement of parasites from one animal to another is more likely. Usually ectoparasites (often obligate ectoparasites) such as lice or mange mites are problems in confinement systems. In fact, lice are a major problem in confined hog and poultry operations. Ectoparasites also are of greater importance in pasture systems during the winter. Another consideration in confined systems is in proper manure management. Many pest species have life histories that depend on the presence of manure. Improper sanitation and manure disposal can contribute to increases in fly populations and increased risk from mechanically transmitted diseases. Pasture or range production systems usually have greater problems with biting flies and ticks. Management of these pests may be complicated by the dispersal of both livestock and pest species over a large area.

OBJECTIVES

1) Understand how pest management of medical and veterinary pests differs from pest management of plant pests.

2) Understand differences between direct and indirect effects of medical and veterinary pests. Also know basic features and types of disease transmission by arthropods.

3) Know major features of life histories and be able to sight identifiy the following medical pest species or groups: black flies, human lice (head, body, and pubic louse), itch mite, and mosquitoes.

4) Know major features of life histories and be able to sight identify the following veterinary pest species or groups: fleas, face fly, horn fly, house fly, livestock lice, mites, sheep ked, stable fly, and ticks.

5) Additional objectives (at instructor's discretion):

OPTIONAL DISPLAYS AND EXERCISES

1) Select a disease, vector, and host from Table 8.2, and learn details of disease transmission.

2) Learn the major pests, production procedures, and recommended pest managment practices for a livestock production system in your area.

3) Examine the displayed photographs or slides indicating illustrating injury by various livestock pests.

4) Examine the displayed groups of insects (which include pests and nonpests) and identify the insect pest.

5) Additional displays or exercises (at instructor's discretion):

TERMS

allergic reaction
annoyance
anthropophilic
biological
   transmission
dermatosis
direct effects
disease
ectoparasite

endemic
endoparasite
entomophobia
envenomation
epidemic
epidemiology
facultative
indirect effects
mange

mechanical
   transmission
medical pest
myiasis
obligate
pandemic
parasite
pathogen
pathology

reservoir
resistance
scabies
synanthropic
urticating hairs
vector
vesicating fluid
veterinary pest
virulence

MAJOR MEDICAL AND VETERINARY PESTS

bed bug	*Cimex lectularius*	Hemiptera: Cimicidae
black flies	Simuliidae spp.	Diptera: Simuliidae
black widow spider	*Latrodectus mactans*	Araneida: Theridiidae
brown recluse spider	*Loxosceles reclusa*	Araneida: Loxoscelidae
cattle grubs:		
common cattle grub	*Hypoderma lineatum*	Diptera: Oestridae
northern cattle grub	*Hypoderma bovis*	Diptera: Oestridae
chiggers	Trombiculidae spp.	Acari: Trombiculidae
esp. common chigger	*Trombicula alfreddugesi*	Acari: Trombiculidae
conenose bugs	Triatominae spp.	Hemiptera: Reduviidae
= kissing bugs		
face fly	*Musca autumnalis*	Diptera: Muscidae
fleas:		
cat flea	*Ctenocephalides felis*	Siphonaptera: Pulicidae
dog flea	*Ctenocephalides canis*	Siphonaptera: Pulicidae
horn fly	*Haematobia irritans*	Diptera: Muscidae
horse and deer flies	Tabanidae spp.	Diptera: Tabanidae
horse bot flies:		
horse bot fly	*Gasterophilus intestinalis*	Diptera: Gasterophilidae
nose bot fly	*Gasterophilus haemorrhoidalis*	Diptera: Gasterophilidae
throat bot fly	*Gasterophilus nasalis*	Diptera: Gasterophilidae
house fly	*Musca domestica*	Diptera: Muscidae
human lice:		
body louse	*Pediculus humanus humanus*	Anoplura: Pediculidae
crab louse	*Pthirus pubis*	Anoplura: Pthiridae
=pubic louse		
head louse	*Pediculus humanus capitis*	Anoplura: Pediculidae
livestock lice:		
cattle biting louse	*Bovicola bovis*	Mallophaga: Trichodectidae
cattle tail louse	*Haematopinus quadripertusus*	Anoplura: Haematopinidae
chicken body louse	*Menacanthus stramineus*	Mallophaga: Menoponidae
hog louse	*Haematopinus suis*	Anoplura: Haematopinidae
horse biting louse	*Bovicola equi*	Mallophaga: Trichodectidae
horse sucking louse	*Haematopinus asini*	Anoplura: Haematopinidae
longnosed cattle louse	*Linognathus vituli*	Anoplura: Linognathidae
sheep biting louse	*Bovicola ovis*	Mallophaga: Trichodectidae
sheep body louse	*Linognathus ovallus*	Anoplura: Linognathidae
shortnosed cattle louse	*Haematopinus eurysternus*	Anoplura: Haematopinidae
mites:		
chicken mite	*Dermanyssus gallinae*	Acari: Dermanyssidae
itch mite	*Sarcoptes scabiei*	Acari: Sarcoptidae
northern fowl mite	*Ornithonyssus sylviarum*	Acari: Macronyssidae
notoedric mange mite	*Notoedres cati*	Acari: Sarcoptidae
sheep scab mite	*Psoroptes ovis*	Acari: Psoroptidae

mosquitoes	Culicidae spp.	Diptera: Culicidae
punkies	Ceratopogonidae spp.	Diptera: Ceratopogonidae
screwworm	*Cochliomyia hominivorax*	Diptera: Calliphoridae
sheep bot fly	*Oestrus ovis*	Diptera: Oestridae
sheep ked	*Melophagus ovinus*	Diptera: Hippoboscidae
stable fly	*Stomoxys calcitrans*	Diptera: Muscidae
stinging Hymenoptera:		
ants	Formicidae spp.	Hymenoptera: Formicidae
bumble bees	Bombidae spp.	Hymenoptera: Bombidae
honey bee	*Apis mellifera*	Hymenoptera: Apidae
hornets/wasps/yellowjackets		
	Vespidae spp.	Hymenoptera: Vespidae
ticks:		
American dog tick	*Dermacentor variabilis*	Acari: Ixodidae
brown dog tick	*Rhipicephalus sanguineus*	Acari: Ixodidae
ear tick	*Otobius megnini*	Acari: Argasidae
deer tick	*Ixodes dammini*	Acari: Ixodidae
Gulf Coast tick	*Amblyomma maculatum*	Acari: Ixodidae
lone star tick	*Amblyomma americanum*	Acari: Ixodidae
Rocky Mountain wood tick		
	Dermacentor andersoni	Acari: Ixodidae
wool maggots	Calliphoridae spp., esp.:	Diptera: Calliphoridae
secondary screwworm	*Cochliomyia macellaria,*	Diptera: Calliphoridae
	Phaenicia sericata,	Diptera: Calliphoridae
black blow fly	*Phormia regina,* &	Diptera: Calliphoridae
	Protophormia terraenovae	Diptera: Calliphoridae

INSECT PESTS BY HOST/PRODUCTION SYSTEM

Beef Cattle - range and pasture

cattle grubs (common cattle grub and northern cattle grub), cattle lice (cattle biting louse, cattle tail louse, longnosed cattle louse, and shortnosed cattle louse), face fly, horn fly, horse flies, mites (sheep scab mite), mosquitoes, stable fly, and ticks (American dog tick, Gulf Coast tick, lone star tick, Rocky Mountain wood tick, and ear tick)

Beef Cattle - confined

cattle grubs (common cattle grub and northern cattle grub), cattle lice (cattle biting louse, cattle tail louse, longnosed cattle louse, and shortnosed cattle louse), house fly, mites (itch mite, sheep scab mite), and stable fly

Dairy Cattle

cattle grubs (common cattle grub and northern cattle grub), cattle lice (cattle biting louse and longnosed cattle louse), face fly, horn fly, horse flies, house fly, mites (itch mite), mosquitoes, and stable fly

Horses

horn fly, horse bots (horse bot fly, nose bot fly, throat bot fly), horse flies, lice (horse biting louse and horse sucking louse), mites (itch mite), stable fly, and ticks

Medical

>bed bug, blackflies, black widow spider, brown recluse spider, chiggers, conenose bugs, horse and deer flies, human lice (body louse, head louse, and pubic louse), itch mite, mosquitoes, punkies, stinging Hymenoptera, and ticks

Pets (Cats and Dogs)

>fleas (cat flea and dog flea), mites (itch mite and notoedric mange mite), and ticks (American dog tick, brown dog tick, and spinose ear tick)

Poultry

>bed bug, house fly, lice (chicken body louse), mites (chicken mite, and northern fowl mite), and mosquitoes

Sheep

>black flies, horse and deer flies, house fly, lice (sheep biting louse and sheep body louse), mites (sheep scab mite), mosquitoes, sheep bot fly, sheep ked, ticks (Gulf Coast tick, lone star tick, Rocky Mountain wood tick, and ear tick), and wool maggots

Swine

>house fly, lice (hog louse), mites (itch mite), mosquitoes, stable fly

DISCUSSION AND STUDY QUESTIONS

1) How might pest management for a mosquito species vectoring a human disease be different from management of a mosquito species that bites humans but does not vector any diseases?

2) What behavioral and life history features of the house fly make it a good vector for mechanically transmitting disease?

3) Using your knowledge of flea biology, what recommendations would you make for flea control on cats or dogs in a household? In a commercial kennel?

4) Why might the trend toward larger poultry rearing facilities offer greater potential for pest problems?

5) What problems might be posed by managing cattle pests on rangeland cattle versus dairy cattle?

BIBLIOGRAPHY

General

Harwood, R. F., and M. T. James. 1979. Entomology in human and animal health. Macmillan Pub. Co., Inc., New York, NY
- an excellent book on medical and veterinary entomology, with emphasis on medical pests. It provides thorough coverage of major groups of medical pests with considerable information on arthropod-borne diseases. This is a readable, and extremely useful text.

Kettle, D. S. 1987. Medical and veterinary entomology. John Wiley and Sons, New York, NY
  - another good introductory text on medical and veterinary entomology.

Williams, R. E., R. D. Hall, A. B. Broce, and P. J. Scholl, eds. 1985. Livestock entomology. John Wiley and Sons, New York, NY
  - a good textbook on livestock entomology that covers all major groups of livestock pests. Considerable information on individual pest species is provided.

History

McNeil, W. H. 1976. Plagues and people. Anchor Press/Doubleday, Garden City, NY
  - an insightful treatment of the role of disease in human history by one of America's preeminent historians. A fascinating book, particularly for its ecological perspective into many historical questions.

Tuchman, B. W. 1978. A distant mirror. A. A. Knopf, New York, NY
  - although entire books have been written on the plague pandemic of the Middle Ages, Barbara Tuchman's account of the late 1300's places the disease in perspective as one feature of a tumultous era. This book provides a superior account of how the Europe responded to the plague. In our opinion one of the finest histories written this century.

Zinsser, H. 1934. Rats, lice, and history. Little, Brown, and Co., New York, NY
  - as Zinsser puts it, this is the biography of a disease. A delightful, witty book.

Epidemiology

Crichton, M. 1969. The andromeda strain. A. A. Knopf, New York, NY
Fuller, J. G. 1974. Fever! The hunt for a new killer virus. Reader's Digest Press, New York, NY
Shilts, R. 1987. And the band played on. St. Martin's Press, New York, NY
  - these three books (the first science fiction, the other two nonfiction) give an exciting, and frightening picture of infectious disease. Although none directly address insect-vectored diseases, most of the procedures and processes they illustrate apply as equally to arthropod-borne diseases as to others. We recommend reading these in chronological order, as the earlier books (including Crichton's fictional story) tend to foreshadow later events. All three books are tremendously exciting - scientific thrillers. After reading them, particularly Shilts's book on AIDS, it is easier to appreciate how people feared and faced plagues and epidemics before antibiotics, vaccines, and the germ theory of disease.

9 STORED PRODUCT PESTS

Grain in storage represents a massive, concentrated food supply for a variety of insect pests. The elevators, bins, mills, warehouses, and wholesale and retail outlets in which grain and other foodstuffs are stored represent suitable ecosystems for supporting large numbers of damaging insects. Since these ecosystems are often free of natural enemies, and because the food supply is virtually inexhaustable, undetected insect populations may increase and expand to explosive levels in a very short time.

Stored product pests not only consume grain and other foodstuffs, but also may render these products useless by defiling them with webbing, frass, exuviae, and whole or fragmented carcasses. Also, the activities of these pests in grain may cause heating of the stored product. The heating induces moisture-laden warmed air to rise to the surface of the grain where it meets cooler air. The collision of the two air masses causes condensation of the accumulated moisture onto the surface of the stored grain. Wet grain encourages mold growth and subsequent spoilage.

The importance of insect damage to foodstuffs during assembly, storage, processing, packaging, and selling should not be underestimated. The USDA Agricultural Research Service estimates that stored product insects cause the direct loss of nearly 40 million tons of food each year in the US. Additional indirect losses to the food industry are incurred due to: (1) the costs of employing insect pest management tactics (e.g., fumigation with insecticides), (2) dockage, consumer complaints, and loss-of-sales due to insect contamination of foodstuffs, and (3) the costs that accompany the recalls, seizures, prosecutions, and injunctions that may result from the sale and distribution of food products containing insects or insect parts.

TRAITS OF STORED PRODUCT PESTS

Many types of insects infest grain and other stored food materials, but they share some common traits. Nearly all of the insects classified as stored product pests are **cosmopolitan** in distribution, that is, they occur throughout the world. Original sources for these insects were the tropical and subtropical regions. Distribution to other areas was accomplished swiftly and early in human history due to mankind's tendency to travel and trade. The tropical origins of the stored grain insects suggest that warm conditions are best for their survival and reproduction. This is true; with only a few minor exceptions, these insects do not live long at low temperatures and do not have a winter diapause or hibernation period. Stored grain insects also are all well-adapted to feeding on and breeding in stored grains and other foods of relatively low moistures. Temperature and moisture level of the storage environment therefore are two primary considerations in managing insect populations in stored products.

While some traits are shared among all of the stored product insects, there are also some major differences in feeding habits that allow us to group these pests into four convenient categories. **Internal feeders** are the insects whose larvae feed and develop completely within a kernel of grain. Internal feeders include the Angoumois grain moth (Fig. 9.1A), granary weevil (Fig. 9.1B), and rice weevil.

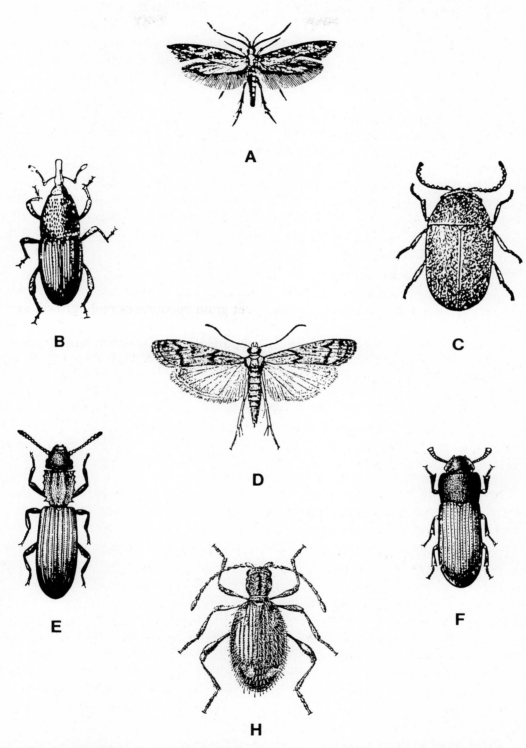

Figure 9.1   Some important stored product pests, including internal feeders (A & B), external feeders (C & D), scavengers (E - G), and secondary pests (H). A. Angoumois grain moth. B. Granary weevil. C. Cigarette beetle. D. Indianmeal moth. E. Confused flour beetle. F. Mediterranean flour moth. G. Sawtoothed grain beetle. H. Spider beetles. (A, B, D, E, G Courtesy Kansas State University; C, F, H Courtesy USDA)

External feeders are insects that feed from the outside of the grain kernels, although they may chew through the outer seed coat and then devour the entire internal contents of the kernel. The cadelle, cigarette beetle (Fig. 9.1C), drugstore beetle, Indianmeal moth (Fig. 9.1D), and lesser grain borer are considered external feeders (although the lesser grain borer, with its aggressive boring activity, is sometimes placed in the internal-feeders category). External feeders frequently are found on other stored foodstuffs such as cereals, nuts, seeds, tobacco, and spices.

The **scavenger** category includes those insects that feed on grain only following breakage of the seed coat by mechanical means or after feeding by internal or external feeders. Scavengers also commonly infest milled products and processed foods of many kinds. Scavengers include the confused and red flour beetles (Fig. 9.1E), flat and foreign grain beetles, Mediterranean flour moth (Fig. 9.1F), and sawtoothed grain beetle (Fig. 9.1G).

A **secondary pest** is a stored product insect that feeds only on foodstuffs that are damp, out of condition, and have some mold growth. Some of these insects feed primarily on the mold rather than the commodity. Many secondary pests are principally nuisances but do serve as indicators of damaged, wet, or decaying stored goods. Examples of secondary pests are the lesser and yellow mealworms, and the spider beetles (Fig. 9.1H).

MANAGEMENT TACTICS ON THE FARM

Proper identification and categorization of stored product pests and knowledge of the insects' behavioral and biological characteristics are integral to the proper employment of pest management tactics. For example, an insecticide treatment incorporated into the surface portion of a grain mass in an on-farm storage bin might control an Indianmeal moth infestation at the upper levels of the grain but will not control a weevil problem that exists in the core of a grain mass. Likewise, cleaning grain to remove fine material will contribute greatly to eliminating some scavengers such as the red flour beetle, but this action is of only limited value for reducing populations of internal feeders such as the rice weevil.

Prevention of insect infestations in foodstuffs is the highest priority in any stored product insect management scheme. Prevention is important for a number of reasons. First, insect-infested grain may be substantially docked at sale time, and some buyers may even refuse to accept infested shipments that might contaminate their storage or processing facilities. Also, insects in farm-stored grain may affect the grain's eligibility for government grain reserve programs. Additionally, foreign grain buyers, upon whom US producers increasingly depend, frequently demand high quality, insect-free grain. Finally, contamination or adulteration of food or feed can be a violation of the **Food, Drug, and Cosmetic Act**. This federal act was passed in 1906 and is enforced by the US Food and Drug Administration (FDA). The act is a broad law with many provisions, but its implications are fairly simple: sanitation in food processing and storage sites must be maintained at high levels, and efforts must be made to **prevent** insect infestaion of foodstuffs. This is because the FDA considers a product contaminated whether the insects infesting it are alive or dead. Similarly, insect fragments are as undesirable as whole insects. Also, insecticide residues in foodstuffs are restricted to quite low levels. For this reason, volatile fumigants are often the sole means of control once insects have infested grain and foods; very few fumigants are currently available for use on foodstuffs and these tend to be rather expensive and dangerous to employ. Prevention of insect infestation of foodstuffs obviously must be the first line of defense in any stored product ecosystem where tolerance for insects is low.

Stored product pest prevention begins at the farm level and, ideally, should not end until the food product is used by the consumer. On the farm, sanitation is the simplest and most important weapon for inhibiting stored product pest infestations. The most common source of a grain pest infestation is old grain in combines, truckbeds, augers, bins, or anywhere else that grain passes through or is stored. Prior to harvesting, all these areas should be thoroughly cleaned. As an added precaution, it is wise to feed the first several bushels of grain going through a combine to livestock. This practice insures a fairly complete flushing of any old, leftover grain from the implement. Before placing new grain in storage, it is also important for the farmer to remove any piles of spilled grain near storage bins that might serve as a source for new insect infestations.

Another on-farm preventive measure is application of a registered insecticide to the walls and floors of empty storage bins to supplement sanitation efforts. These **prebinning insecticide** residues provide control of insects that may remain in hard-to-clean cracks and crevices in the storage areas. Prevention also may include the application of **protectant** insecticides directly to newly-harvested grain. Insecticides labeled for this purpose frequently are applied to grain as it flows into an auger, conveyor belt, or other transfer equipment. The stability of protectant insecticides is largely determined by grain temperature and moisture, with low temperatures and low moisture favored for long-term insecticide effectiveness. Of course, it is always important for farmers to engage proper drying, aeration, and temperature regulation practices for grain in storage. Moist, poorly-aerated grain is far more subject to attack by insects and storage fungi than is well-maintained grain.

Although on-farm prevention of insect infestation is the best management tactic, sometimes it is necessary to control existing infestations. Techniques used include aerating to cool the grain to below 50° F to prevent insect activity and allow an extended period of storage. This tactic tends to slow development of an infestation but generally will not eliminate it. When an infestation is limited to a small portion of a grain mass, as occurs with some Indian-meal moth infestations, a spot treatment with registered and recommended insecticides incorporated into the infested grain may be useful. When internal feeders such as the rice weevil have infested grain in storage, fumigation is often necessary to kill the egg, larval, and pupal stages that occur within the kernel. **Fumigants** are pesticides that exist in the gas phase at effective temperatures, as compared to fogs or aerosols which are dispersions of very fine particles or droplets. Fumigants penetrate cracks, crevices, and the commodity being treated. As soon as the gas diffuses from the area its effectiveness is lost (virtually no residual activity) and reinfestation may occur. Fumigants must only be applied in enclosed areas to be effective. Only highly-trained applicators, equipped with specialized protective equipment, may apply most fumigants. These procedures are necessary because most fumigants are highly toxic, and some are very flammable and explosive.

Once grain is in storage, setting up a regimented monitoring system for the entire storage period is imperative. Monitoring is necessary for detecting potential or developing insect problems and for ensuring that prompt emergency treatments can be instituted should problems arise. Grain in storage should be regularly examined for signs of crusting, moist areas, moldy spots and musty odors, temperature changes, and actual insects or insect products (such as webbing). Various probes, temperature sensors, and screening devices are commercially-available for conducting the necessary inspections. Additionally, various traps are sold for detecting specific insects in grain storage areas. These traps are typically baited with synthetic sex pheromones specific to the insect of interest. A **sex pheromone** is an intraspecific chemical signal produced by one sex for the purpose of mate location.

MANAGEMENT TACTICS IN FOOD HANDLING ESTABLISHMENTS

Food processing and packaging plants and wholesale and retail food outlets must practice many of the same management techniques employed by the farmer. Commercial food-handling establishments must, however, comply very carefully with the high standards of pest control and sanitation expected by the public and enforced by various regulatory agencies, including the US Food and Drug Administration and the Meat and Poultry Inspection Program of the USDA. Naturally, monitoring and prevention are the first lines of defense against stored-product insect infestation. While each food industry has its own special quality control standards, some general tactics are employed to prevent and manage pests in food storage and production facilities.

Food handling facilities should be tightly constructed so that dirt and debris do not accumulate and so insects do not have places to live, feed, and breed. Walls, floors, and ceilings should be constructed of materials that are impervious and therefore easy to clean and treat with insecticides when necessary and permissible. Incoming deliveries of foodstuffs and other stock should be thoroughly inspected, and infested materials should be refused. Pre-palletized deliveries should be discouraged as they inhibit proper inspection of individual cartons or sacks of goods. Regular and thorough inspection and monitoring of storage areas for signs and symptoms of infestations are of great importance. Raw food materials should be stored in areas separate from finished goods, all foodstuffs should be stored in tightly sealed containers, and all foods should be protected from moisture and other damage. Stored goods should always be held above the ground on racks or pallets and clear aisles should surround the raised stock. Stock rotation (using oldest materials first) should be carefully practiced to prevent the accumulation of old and possibly spoiled or infested goods. Equipment used for processing or packaging foods must be kept scrupulously clean. Any surfaces which come in direct contact with the food product must be made of smooth, nonporous materials such as stainless steel. All areas inside equipment should be readily accessible for cleaning so there are no hidden corners or crevices to accumulate debris and harbor insects.

The use of pesticides in and around food handling establishments is highly restricted as a result of amendments to the Federal Insecticide, Fungicide, and Rodenticide Act (FIFRA). In most food areas, application of residual pesticides is limited to placement in unexposed cracks and crevices such as openings at expansion joints, between equipment and floors, and between different elements of construction. The treatment of surfaces is not allowed, and every effort must be made to prevent insecticide residues from being left on a surface that may come in direct contact with food or food products. In nonfood areas of food handling establishments, slightly more liberal application of residual pesticides to surfaces is permitted, but pesticides must not be allowed to come in contact with food containers of any kind. Any pesticide application in a food handling establishment must be conducted in strict accordance with existing laws, according to instructions on the pesticide labels, and with genuine awareness of, and respect for, a simple fact: no insecticide should be considered as "nontoxic" or suitable for use directly on any food product.

OBJECTIVES

1) Understand how insect pest management in stored-product ecosystems differs from management in agroecosystems.

2) Recognize common pest management tactics used in grain storage and food handling systems.

3) Know major features of life histories and be able to sight identify the following insects: granary weevil, rice weevil, cadelle, Indianmeal moth, confused flour beetle, red flour beetle, sawtoothed grain beetle, yellow mealworm.

4) Understand the various feeding categories of stored product insects and know which category each of the above insects falls into.

5) Additional objectives (at instructor's discretion):

OPTIONAL DISPLAYS AND EXERCISES

1) Examine the displayed cultures of stored product insects.

2) Determine which stored product fumigants are currently available for use in your area. How are they applied? What skills, special training, or special equipment are needed for their application? What special dangers are posed by each of the fumigants?

3) Examine the displayed examples of tools and protective equipment used in fumigation procedures.

4) Examine the displayed examples of sampling and surveilance equipment used to monitor stored grain for insect infestation.

5) Additional displays or exercises (at instructor's discretion):

TERMS

cosmopolitan
Food, Drug, and
   Cosmetic Act

external feeder
fumigant
grain protectant

internal feeder
prebinning
  insecticide

scavenger
secondary pest
sex pheromone

MAJOR STORED PRODUCT PESTS

Angoumois grain moth	*Sitotroga cerealella*	Lepidoptera: Gelechiidae
cadelle	*Tenebroides mauritanicus*	Coleoptera: Trogositidae
cigarette beetle	*Lasioderma serricorne*	Coleoptera: Anobiidae
confused flour beetle	*Tribolium confusum*	Coleoptera: Tenebrionidae
drugstore beetle	*Stegobium paniceum*	Coleoptera: Anobiidae
flat grain beetle	*Cryptolestes pusillus*	Coleoptera: Cucujidae
foreign grain beetle	*Ahasverus advena*	Coleoptera: Cucujidae
granary weevil	*Sitophilus granarius*	Coleoptera: Curculionidae
Indianmeal moth	*Plodia interpunctella*	Lepidoptera: Pyralidae
lesser grain borer	*Rhyzopertha dominica*	Coleoptera: Bostrichidae
lesser mealworm	*Alphitobius diaperinus*	Coleoptera: Tenebrionidae
Mediterranean flour moth	*Anagasta kuehniella*	Lepidoptera: Pyralidae
red flour beetle	*Tribolium castaneum*	Coleoptera: Tenebrionidae
rice weevil	*Sitophilus oryzae*	Coleoptera: Curculionidae
sawtoothed grain beetle	*Oryzaephilus surinamensis*	Coleoptera: Cucujidae
spider beetles	Ptinidae spp.	Coleoptera: Ptinidae
yellow mealworm	*Tenebrio molitor*	Coleoptera: Tenebrionidae

INSECT PESTS BY TYPE OF INJURY

Internal Feeders
    Angoumois grain moth, granary weevil, and rice weevil
External Feeders
    cadelle, cigarette beetle, drugstore beetle, Indianmeal moth, and lesser grain borer
Scavengers
    confused flour beetle, flat grain beetle, foreign grain beetle, Mediterranean flour moth, red flour beetle, and sawtoothed grain beetle
Secondary Pests
    lesser mealworm, spider beetles, and yellow mealworm

DISCUSSION AND STUDY QUESTIONS

1) What factors should be considered when designing a pest management scheme for dealing with stored-product-insect infestations in a home pantry? A grocery warehouse? A railcar full of raw grain?

2) What are the advantages and disadvantages of using sex pheromone-baited traps for detecting and monitoring stored product insect infestations?

3) What is the feasibility of using biological control tactics in stored product insect pest management?

4) Methyl bromide is an odorless, colorless, and tasteless fumigant that is 3.3 times heavier

than air, and it reacts with rubber. Knowing these facts, what special precautions would you think necessary for safe handling of this material?

5) A hybrid corn seed storage facility exhibits a severe infestation of Angoumois grain moth. Speculate on the suitability of the following management tactics: surface treatment of the stored seed with *Bacillus thuringiensis* (a stomach poison for lepidopteran larvae); freezing of the seed; fumigation with methyl bromide or aluminum phosphide; fumigation with CO_2.

6) What are the advantages and disadvantages of the following types of food packaging materials from a pest management standpoint: glass, steel, cellophane, paper, and cloth?

BIBLIOGRAPHY

Agricultural Research Service, USDA. 1986. Stored-grain insects. USDA-ARS Handbook Number 500
- a small handbook that describes the biology, life histories, and habits of more than 50 insects associated with stored foods. Contains some photographs and line drawings.

American Cyanamid Company. 1983. How to control insects in farm-stored grain. PE4016
- a comprehensive management guide designed for the grower. Naturally, use of American Cyanamid insecticides is suggested, but the guide also contains much useful information on noninsecticidal management tactics.

Baur, F. J., ed. 1984. Insect management for food storage and processing. American Association of Cereal Chemists, St. Paul, MN
- a multiauthor publication covering many aspects of insect pest management in the consumer products industries.

Gorham, R. J., ed. 1981. Principles of food analysis for filth, decomposition, and foreign matter. US Food and Drug Administration Technical Bulletin No. 1
- a facinating journey through the world of the "analytical entomologist". Gives detailed description of the many arthropod and nonarthropod impurities that may be found in foods. Also covers sanitation and analytical techniques and tools employed in the evaluation of food purity.

Journal of Stored Product Research. Pergamon Press, New York, NY
- a quarterly-issued journal which publishes original research papers "dealing with the biology, ecology, physiology, behavior, taxonomy, genetics, or control of the insects, mites, fungi, and other organisms associated with stored products or farm and domestic buildings, or with studies on the physical and chemical nature of the environment which have relevance to these organisms and their control."

Mallis, A. 1982. Handbook of pest control. 6th ed. Franzak & Foster Co., Cleveland, OH
- this book contains a comprehensive chapter on common and rare stored product pests and their behaviors, life histories, and control. It has a number of color plates and line drawings.

Munro, J. W. 1966. Pests of stored products. Hutchinson Co., Ltd., New York, NY
   - although this book is slightly dated, it contains an abundance of information of stored product pest identification and biology. It contains many detailed drawings of insects and some photographs.

Peace, D. McC. 1985. Key for the identification of mandibles of stored-food insects. Association of Official Analytical Chemists, Arlington, VA
   - this publication is designed for use by quality control personnel in the food industries. Since fragments often must be used to identify insects in food, this key has real practical utility (mandibles are the favored fragments for identification because of their hardness, darkness, uniqueness within a species, and because they often remain intact when the rest of the insect body is destroyed). Most of the drawings in this key are large and fairly detailed.

Truman, L. C., G. W. Bennett, & W. L. Butts. 1976. Scientific guide to pest control operations. Purdue University/Harvest Publishing, Cleveland.
   - contains an information-packed chapter on stored product pests designed for workers in the urban pest control industry. Contains a number of photos, line drawings, and keys.

Wilson, M. C., G. W. Bennett, & A. V. Provonsha. 1977. Practical insect pest management. Vol. 5. Insects of man's household & health. Waveland Press, Inc., Prospect Heights, IL.
   - A course manual in a series on practical insect pest management. Contains a chapter on "Insects Infesting Stored Products".

10 URBAN PESTS

Urbanization limits our contact with many types of wildlife but often serves to intensify our contact with others. The urban environment provides many ideal habitats for pests that are well-adapted for survival and reproduction in our buildings, stored goods, and refuse. Certain characteristics enhance the suitability of urban ecosystems as pest habitats. They tend to be people-dense, and thus, resource-rich. Seldom does any human structure lack the food, moisture, or shelter required for pest establishment and maintenance. Urban ecosystems also tend to be intensely interconnected. The constant movement of goods and services from structure to structure allows for rapid and extensive distribution of pests to new resources. Finally, urban ecosystems are complex and multifaceted. A single structure may provide habitats for a vast array of pests: moist wood around a small roof leak may provide a carpenter ant nest site, the dark and damp area around the basement drain may give shelter to oriental cockroaches, and the accumulation of crumbs and grease between the stove and refrigerator may act as a veritable smorgasbord for German cockroaches, house flies, and Pharaoh ants.

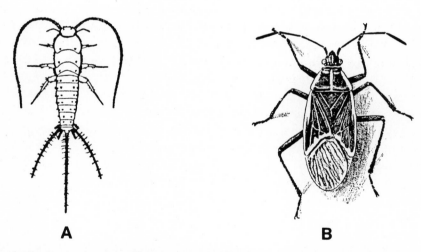

A   **B**

Figure 10.1 Types of urban pests. A. Silverfish, a synanthropic insect. B. Boxelder bug, an accidental invader. (Courtesy USDA)

TYPES OF URBAN PESTS

Urban pests may be conveniently grouped into two main categories: (1) structure-destroying pests, and (2) structure-invading pests. Structure-destroyers include those organisms which actually feed on or burrow into woods and other construction materials. This category thus encompasses all type of termites, carpenter ants, and many types of wood-boring beetles. Structure-invaders include many **synanthropic** species (those living in close association with humans) which can live indefinitely within the confines of human habitations, and **accidental invaders** which may regularly invade structures but which seldom are capable of

191

reproduction and proliferation indoors. Synanthropic species include pests such as silverfish (Fig. 10.1A) and German cockroaches, whereas accidental invaders include pests such as field crickets and boxelder bugs (Fig. 10.1B). The management approaches taken and the control tactics employed in the urban setting will depend largely on the category of the pest and, as we shall see, the specific urban ecosystem in which the pest occurs.

IMPORTANCE OF URBAN PESTS

The economic, aesthetic, and public health problems posed by urban pests should not be underestimated. The direct and indirect costs of dealing with the insects common to urban environments are considerable and the pest management problems presented by urban pest complexes are formidable. In some instances, urban pests may cause direct and costly damage to stored products (e.g., larder beetle, webbing clothes moth) and to structures (e.g., termites). Losses due to termites alone in the US have been estimated at well over $1 billion annually. More difficult to quantify is the cost of urban pests to the physical and mental well-being of people (e.g., house flies entering a nursing home or cockroaches interrupting and spoiling dinner for a restaurant patron). Such urban pests may carry an array of pathogenic organisms in and on their bodies. Fortunately, however, the inefficiency of mechanical pathogen transmission makes such pests primarily bothersome rather than dangerous.

In all cases, urban pest management differs from agricultural pest management in that the concept of an economic threshold rarely applies in the urban setting: one fly in a customer's soup is one too many, one colony of termites in a neighborhood is one too many, one cockroach in a processed meat product is one too many, one drain fly in a hospital operating room is one too many, *ad infinitum*. People tend to either abhor or fear most urban pests and, therefore, will not tolerate insects, at any population level, in their homes, foods, or personal effects. It is easy to think that the presence of a zero-threshold for most urban pests indicates that integrated pest management approaches have little or no applicability in the urban setting. On the contrary, in situations where the presence of certain pests is highly objectionable, the intelligent use of pest control options in an economically, ecologically, and sociologically sound manner is an absolute necessity. Proper inspection, identification, treatment, and monitoring of pest problems in the urban setting is necessary, if the absence or near absence of pests in a particular urban ecosystem is desired or required.

PEST MANAGEMENT IN A FOOD-SERVICE OPERATION

To illustrate the numerous dilemmas encountered in urban pest management, we will examine problems frequently associated with two common urban ecosystems: a food-service operation (for instance, a small restaurant) and a multifamily dwelling (an apartment building). A food-service operation's two primary pest problems are usually house flies and German cockroaches.

The house fly problem often stems from improper garbage disposal in the outdoor areas behind the operation. Sloppy disposal of trash, obsolete receptacles with poorly sealed lids, infrequent garbage pick-up, and inadequate or infrequent cleaning of receptacles often are the primary contributing factors to the problem. Such practices provide a suitable breeding ground for large numbers of flies. As adult flies emerge from the garbage they often congregate around the doors of the food-service operation, attracted by the various food odors emanating from the building. Flies may enter through improper seals around the doors and windows or when doors are opened for deliveries or patrons. Air curtains, fanlike devices that keep insects, dirt, and other airborne materials from entering open doors, often help eliminate fly problems. Unfortunately, the devices often are too expensive for small businesses to install and maintain, and they may increase energy costs.

Once flies have entered the food-service operation control is difficult. The use of fly paper or other types of traps is discouraged by most health departments. Black-light electrocution traps are of some utility but are expensive and not completely effective. Vapona-impregnated strips ("no-pest strips") cannot be used in the vicinity of food, and currently the EPA is considering cancelling all uses of Vapona. Fogging procedures may be performed only when patrons or employees are absent and when all foods and food-preparation surfaces have been protected. The house fly, therefore, may often enter such operations at will and proceed upon its merry way, annoying patrons and employees and distributing vomit and fecal spots as it goes.

The German cockroach problems encountered in food-service operations are generally the result of three factors: (1) continual introduction of cockroaches in deliveries and with patrons, employees, and their belongings; (2) high availability of cockroach **harborage** in various restaurant equipment combined with marginal sanitation in and around harborage sites; and (3) inadequate or improper treatment of existing problems and little or no monitoring for new problems.

If suppliers to a food-service operation have cockroach infestations, then it is quite likely that their products will contain cockroaches or cockroach egg cases as well. Proper inspection of deliveries and rejection of infested shipments are necessities. Unfortunately, these procedures are seldom routinely practiced, therefore, a constant influx of new pests can occur. Cockroaches carried on people or, particularly, on their belongings can be another source of reinfestations. Avoiding these introductions requires **prophylactic** (preventive) treatments with residual pesticidal compounds in or about employee lockers (or wherever employee belongings are kept) and around waiting and seating areas frequented by restaurant patrons.

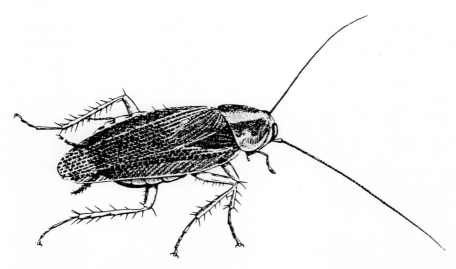

Figure 10.2 German cockroach. (Courtesy Connecticut Agricultural Experiment Station)

Numerous harborage and sanitation problems are encountered in food-service operations. Indeed, to an entomologist restaurant equipment sometimes seems to have been designed specifically for cockroach harborage. Hollow equipment legs, gaps, cracks, and crevices where two or more surfaces meet, and parts that cannot be removed for cleaning provide ideal refuge for the positively **thigmotactic** (contact-loving) German cockroach (Fig. 10.2). Accumulation of damp and greasy bits of food in these areas also benefits the insects. Application of residual pesticides to control cockroaches in these cracks and crevices is possible and desirable, but the moisture and food present in these areas decreases the residual toxicity of many pesticide treatments and may decrease the insects' exposure to them. Ideally, the closure of all gaps, cracks, and crevices to eliminate possible cockroach harborage is the best long-term approach for dealing with the problem. However, this is seldom accomplished due to time and expense and to the lack of understanding by the restaurateur regarding habits and needs of the German cockroach.

Many problems with control of German cockroach infestations are possible. Frequently, the owner of the operation is unwilling to pay for the services of a reputable professional pest control technician or refuses to allow inspection and treatment as frequently as is needed. In other cases, a pest control technician is employed but finds accomplishing his or her tasks difficult; sometimes food-service personnel object to the movement of equipment needed for proper inspection or treatment, find the odor of the pesticides applied to be bothersome, or refuse to cooperate with the pest-control technician in the areas of sanitation and harborage elimination. Of course, there are also instances in which a pest-control technician does not inspect the premise thoroughly and fails to identify and properly treat a problem area. Obviously, it is very important that a high level of communication exist between the food-service personnel and the pest-control technician; they need to inform each other of sites and sources of infestations and must cooperate with one another in terms of finding solutions to the pest problems at hand.

PEST MANAGEMENT IN A MULTIFAMILY DWELLING

A food-service operation and a multifamily dwelling share many urban pest problems but also have many unique characteristics. The multifamily dwelling can be subject to a variety of complex pest problems. The primary factor to keep in mind when assessing pest management in an apartment building is that pest problems are rarely limited to one unit in a multifamily complex; plumbing lines, electrical systems, and ventilation ducts are generally shared among the units and thus may serve as conduits for pest transport from one apartment to another. Another important element of the multifamily dwelling is its heterogeneous make-up; while units within the building may be structurally identical, the residents are unique individuals; each has his or her own ideas about sanitation, food preparation and storage, pesticide exposure, and the acceptability of synanthropic insects. The apartment manager, pest control technician, or other residents can do little to change the ideas and lifestyles of specific residents.

German cockroaches are undoubtedly the most persistent and frustrating pest problem in the multifamily dwelling. Other pests, such as crickets, flies, larder beetles, or silverfish do cause localized problems but, with their more limited mobility, these insects often can be eliminated (e.g., residents can discard food products invested with larder beetles) or their entrance to the apartment can be hindered (e.g., keep doors shut and window screens in place to keep flies and other invaders out).

The management problems involved in controlling cockroaches are numerous. First, some residents fail to report the presence of insects until populations have escalated to overwhelming proportions. By this time, cockroaches often have begun moving into adjacent apartments. Pesticide application in only one apartment obviously will not solve the building's problem, as a fresh source of infestation will be waiting next door. Continuous surveillance (e.g., via the use of sticky traps or "roach motels") to detect new or resurging populations, or even prophylactic pesticide application in vulnerable locations in a multifamily dwelling often is desirable; unfortunately, this is rarely simple and is sometimes impossible. Sometimes an apartment manager is unable to recognize the purpose of monitoring or treating apartments that have not reported infestations; he or she may find it difficult to view the building as a single cockroach ecosystem. Residents, likewise, may offer resistance to having their apartments inspected or treated; some object to the presence of an outsider in their home; some do not want to be troubled with the task of removing items from their cupboards and closets so that the pest control technician can inspect and apply pesticides or other control measures; certain residents fear the presence of pesticide residues in their homes or object to or are sensitive to the odors that frequently accompany pesticide application; yet others refuse to alter their lifestyle through improved sanitation or increased awareness of the insects and the problems associated with them. The pest control technician must therefore often act as an educator of both the manager and residents of the multifamily unit. It is important that all parties recognize the fact that, where cockroaches are concerned, the multifamily dwelling is a single unit affected by the actions of many individual units.

URBAN PEST MANAGEMENT TACTICS

In the previous discussion we alluded to a number of urban pest management tactics. The tactics employed in the urban ecosystem are vast, varied, and constantly changing as the technology of pest control evolves and new management tools become available. However, some general tactics and techniques apply in many situations and always should be included in a four-part pest management plan. The four-part plan is: (1) Inspect. The purpose of the inspection is to identify potential problem areas or sources of infestation and to search for clues regarding the presence and identity of pests. (2) Identify and diagnose. The pests present must be properly identified and the reasons for their infestations should be determined (e.g., poor sanitation, infested deliveries, ripped window screens, etc.). (3) Recommend and take action. Recommendations for preventive measures (e.g., caulking of cracks where cockroaches hide, elimination of refuse where flies breed) should be made in conjunction with application of control tactics. (4) Monitor. Following employment of control tactics, problem and nonproblem areas should be reinspected frequently for the presence of new infestations, changes in environmental conditions (e.g., new water, food, or shelter supplies), or failure of tactics previously employed. Monitoring must continue indefinitely, as the changing urban environment offers pests continual access to a variety of resources.

Specific tactics and techniques differ depending on the nature of the insect pest. Managing structure-destroying pests will be considered first. Termites are among the most destructive insects, and damage prevention is an important management tactic. Because subterranean termites require contact with soil, it is wise to keep all wood out of contact with soil. There should be adequate clearance between soil and house siding, wood porches and steps, or crawl space framing to prevent termites from entering wood. Backfilling with wood debris should be avoided, and fence posts and other wood items that must be in contact with the soil should be treated with creosote or other recommended materials. In areas where termites are a problem a pesticide barrier often is placed between the soil and the foundation of the structure. Pesticides with a long residual life are preferred, but many now have been cancelled because of environmental and health concerns. For some drywood termite infestations fumigation of an entire structure by enclosing it in impermeable tarpaulins under which toxic gases are released is employed. Fumigation of infested wood objects or lumber also is utilized in the management of powder-post beetles and other wood-boring insects. Carpenter ant (Fig. 10.3) infestations may often be prevented by eliminating high-moisture conditions (e.g., a leaky roof) that favor attack by the insect. Treatment of existing infestations involves locating the nest areas and removing infested wood or introducing a residual insecticide into the infested wood.

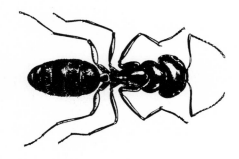

Figure 10.3   Carpenter ant. (Courtesy USDA)

Dealing with synanthropic invaders involves prevention, resource elimination, and application of control measures (e.g., traps, pesticides). Preventing infestations may involve inspection of potentially-infested materials prior to their admittance into a structure. It may also include installation of barriers that restrict entrance into uninfested areas (e.g., sealing holes in screens through which house flies might enter). Resource elimination includes sanitation improvements such as disposing of stacks of paper harboring booklice or silverfish or thorough cleaning of detritus and slime from drains in which drain flies or oriental cockroaches may survive and reproduce. Elimination of harborage is an important part of resource reduction. For instance, sealing of cracks and crevices around sinks and other fixtures can eliminate many German cockroach hiding sites. Proper storage of goods also may inhibit pests from attaining desired resources. For example, woolens sealed tightly in chests and furs placed in cold storage are beyond the reach of clothes moths and other fiber pests.

Control measures used to kill pests directly often involve the use of pesticides. The application of pesticides with residual activity is a common method of dealing with many synanthropic species. For instance, application of organophosphate insecticides to cracks and crevices that harbor German cockroaches is a popular control tactic. Other methods of control include: application of silica gel (a desiccant), boric acid (a stomach poison), insect baits containing pesticides or growth-regulating compounds, fumigants, or aerosols to areas frequented by insects. Occasionally, traps are used for monitoring or controlling insects, especially in areas such as food-processing plants, where the presence of pesticide residues is restricted.

Pests that accidentally invade structures generally are treated as are synanthropic species. Of great importance is exclusion, or preventing such pests from entering structures.

Techniques employed include complete sealing of cracks in foundations, closing of gaps around windows and doors, and repairing holes or tears in window screening. It also may be useful to eliminate outdoor harborage that brings these pests in close proximity to structures. Wood-piles, heavy mulch, dense plantings, or any type of debris should not be allowed in areas immediately adjacent to buildings. Occasionally, application of pesticides with residual activity to the foundation's outdoor perimeter is used to supplement sanitation procedures in areas prone to invasions of pests such as crickets, millipedes, or clover mites.

If accidental invaders do manage to enter a building, manually destroying or vacuuming up stray individuals may be useful. Where an actual "invasion" of many pests occurs, the application of pesticides with residual activity indoors (e.g., around basement window sills or base-boards) may have practical utility. Since most accidental invaders are unable to establish themselves indoors under most circumstances, the home or business owner should be encouraged to view them as a temporary nuisance that does not require a full-scale pesticidal assault.

OBJECTIVES

1) Understand how insect pest management in urban ecosystems differs from management in agroecosystems.

2) Recognize common pest management tactics used for the different categories of urban pests.

3) Know major features or life histories and be able to sight identify the following insects or insect groups: American cockroach, boxelder bug, brownbanded cockroach, carpenter ant, earwig, field cricket, firebrat, German cockroach, house fly, larder beetle, oriental cockroach, silverfish, termites, webbing clothes moth.

4) Additional objectives (at instructor's discretion):

OPTIONAL DISPLAYS AND EXERCISES

1) Select an urban ecosystem and predict pest problems and pest management dilemmas expected in the system (possible ecosystems: hospital, bakery, food processing plant, nursery school, single family dwelling, natural history museum).

2) Identify possible German cockroach harborage in your laboratory/classroom and propose solutions for eliminating harborage.

3) Examine the displayed examples of injury by structure-destroying and synanthropic insects.

4) Examine the displayed examples of tools used by urban pest control technicians.

5) Additional displays or exercises (at instructor's discretion):

TERMS

accidental invader   prophylaxis   syanthropic   thigmotaxis
harborage

MAJOR URBAN PESTS

American cockroach	*Periplaneta americana*	Orthoptera: Blattidae
Asian cockroach	*Blattella asahinai*	Orthoptera: Blattellidae
booklice	various spp., esp.	
booklouse	*Liposcelis corrodens,*	Psocoptera: Lioscelidae
	L. divinatorius,	Psocoptera: Lioscelidae
	& *Trogium pulsatorium,*	Pscoptera: Trogiidae
boxelder bug	*Leptocoris trivittatus*	Hemiptera: Rhopalidae
brownbanded cockroach	*Supella longipalpa*	Orthoptera: Blattellidae
carpenter ants	*Camponotus* spp.	Hymenoptera: Formicidae
carpenter bees	*Xylocopa* spp.	Hymenoptera: Apidae
casemaking clothes moth	*Tinea pellionella*	Lepidoptera: Tineidae
clover mite	*Bryobia praetiosa*	Acari: Tetranychidae
cluster fly	*Pollenia rudis*	Diptera: Calliphoridae
crazy ant	*Paratrechina longicornis*	Hymenoptera: Formicidae
drain and moth flies	Psychodidae spp.	Diptera: Psychodidae
drywood (nonsubterranean) termites		
	Kalotermes spp.	Isoptera: Rhinotermitidae
earwig	Dermaptera spp.	Dermaptera
eastern subterranean termite	*Reticulitermes flavipes*	Isoptera: Rhinotermitidae
false powder post beetles	*Dinoderus, Polycaon,* and *Scobicia* spp.	
		Coleoptera: Bostrichidae
field cricket	*Gryllus* spp.	Orthoptera: Gryllidae
firebrat	*Thermobia domestica*	Thysanura: Lepismatidae
Formosan subterranean termite		
	Coptotermes formosanus	Isoptera: Rhinotermitidae
furniture and deathwatch beetles		
	Anobium, Xestobium, and *Xyletinus* spp.	
		Coleoptera: Anobiidae
German cockroach	*Blattella germanica*	Orthoptera: Blattellidae
house centipede	*Scutigera coleoptrata*	Scutigeromorpha: Scutigeridae
house fly	*Musca domestica*	Diptera: Muscidae
larder beetle	*Dermestes lardarius*	Coleoptera: Dermestidae
millipede	Diplopoda spp.	Class Diplopoda
odorous house ant	*Tapinoma sessile*	Hymenoptera: Formicidae
oriental cockroach	*Blatta orientalis*	Orthoptera: Blattidae
Pharaoh ant	*Monomorium pharaonis*	Hymenoptera: Formicidae

silverfish	*Lepisma saccharina*	Thysanura: Lepismatidae
true powder post beetles	*Lyctus* and *Trogoxylon* spp.	Coleoptera: Lyctidae
webbing clothes moth	*Tineola bisselliella*	Lepidoptera: Tineidae
wood cockroaches	*Parcoblatta* spp.	Orthoptera: Blattellidae

INSECT PESTS BY INJURY CATEGORY

Structure-Destroyers

carpenter ants, carpenter bees, drywood (nonsubterranean) termite, eastern subterranean termite, false powder post beetle, Formosan subterranean termite, furniture and deathwatch beetles, and true powder post beetles

Structure-Invaders (that may reproduce indoors)

American cockroach, Asian cockroach, booklice, brownbanded cockroach, casemaking clothes moth, crazy ant, drain and moth flies, firebrat, German cockroach, house centipede, house fly, larder beetle, odorous house ant, oriental cockroach, Pharaoh ant, silverfish, and webbing clothes moth

Structure-Invaders (that do not reproduce indoors)

boxelder bug, clover mite, cluster fly, earwig, field cricket, house centipede, and wood cockroaches

DISCUSSION AND STUDY QUESTIONS

1) What is the feasibility of using biological control tactics in urban pest management?

2) What factors should be considered in terms of pest management when designing a hospital patient room? A food-storage warehouse? Equipment for use in a food-processing plant?

3) Why will the use of sticky traps (or "roach motels") generally fail to adequately manage a large infestation of German cockroaches in a kitchen?

4) Sometimes people attempt to bring a large German cockroach infestation under control through the use of a total-release aerosol "bomb". These "bombs" generally contain the active ingredient pyrethrins, a stimulatory botanical insecticide with no significant residual activity. Why does this method often fail to adversely affect the cockroach population? Under what circumstances would such a control tactic be of utility?

5) What precautions need to be taken in terms of subterranean termite prevention when a homeowner is constructing an addition to his or her home?

6) Many folk remedies purport to eliminate insect pests from structures. Commonly, it is suggested that residents place items such as sticks of spearmint chewing gum, bay leaves, or fruits of the Osage orange tree ("hedge apples") in basements or closets to either repel or kill pests. Why might such methods appear to succeed or fail in eliminating accidental invaders?

7) Because of your knowledge of entomology and pest management, you have been delegated to educate residents of your apartment building on urban pests and their prevention and

control. What means will you use to achieve this? What areas of pest management will you concentrate on?

BIBLIOGRAPHY

Bennett, G. W. and J. M. Owens, eds. 1986. Advances in urban pest management. Van Nostrand Reinold Co., New York, NY
- a multiauthor collection of chapters that presents "the latest technical developments and the most recent thoughts and principles of urban pest management."

Cornwell, P. B. 1973. Pest control in buildings. Hutchinson of London, London
- a guide to the meaning of terms regarding urban pests, their biology, aspects of public health, pesticides and their application. In glossary form, the book also contains many color photographs.

Cornwell, P. B. 1968. The cockroach. Vol. 1. Hutchinson of London, London
- everything you would want to know about cockroaches: biology, physiology, behavior, and ecology.

Kofoid, C. A., ed. 1934. Termites and termite control. University of California Press, Berkeley, CA
- although this is an older text it contains fairly complete accounts of the biology of the economically-important termite species. While much of the information regarding prevention of termite infestation through proper construction is still valuable and valid, the discussions of pesticidal control are rather out-of-date and obsolete.

Mallis, A. 1982. Handbook of pest control. 6th ed. Franzak and Foster Co., Cleveland, OH
- comprehensive and complete accounts of the behaviors, life histories, and control of all common and many less common household and urban pests. Contains a number of color photographs and many interesting anecdotes. Undoubtedly one of the most significant works on this broad subject.

Truman, L. C., G. W. Bennett, and W. L. Butts. 1976. Scientific guide to pest control operations. Purdue University/Harvest Publishing, Cleveland, OH
- a thorough coverage of urban pest management that is designed for workers in the urban pest control industry. Contains a number of color photographs, diagrams, and keys to pest identification.

Wilson, M. C., G. W. Bennett, and A. V. Provonsha. 1977. Practical insect pest management. Vol. 5. Insects of man's household and health. Waveland Press, Inc., Prospect Heights, IL
- a course manual in a series of manuals on practical insect pest management. Contains a number of excellent illustrations and diagrams but, since it is really just an outline for study, lacks extensive discussion of specific topics.

Periodicals and Trade Magazines

Pest Management. Published by the National Pest Control Association. ARC Publications, Alexandria, VA
- the "voice of the industry".

Pest Control. Harcourt Brace Jovanovich Publications, Cleveland, OH

Pest Control Technology. Gie, Inc. Publishing, Cleveland, OH

SECTION III. MANAGEMENT

The last five chapters address topics related to managing insect pests. The previous chapters primarily focused on the first step in curing an insect pest problem - identifying the insect. Beyond recognizing the insect pest, however, sampling and decision making are central to any pest management program. Consequently, Chapter 11, Sampling and Decision Making, addresses the next two steps in attacking an insect pest problem - quantifying pest numbers and determining if management is warranted.

The last step in managing insect pest problems is utilizing the appropriate pest management tactics, and Chapters 12-15 provide information on important tactics. Specifically, these tactics are insecticidal control, biological control, genetic control, ecological management, plant resistance, and regulatory management.

The intent in the management chapters is to provide background information on sampling, decision making, and individual tactics. Although you may not be directly involved in certain types of pest management tactics, such as genetic control or regulatory management, you need to recognize the array of tactics that may be employed against a pest. Because insecticides are so widely used in pest management and because of the hazards that can be associated with their use, more information is provided on insecticidal control than on other tactics. The material on safe use of insecticides and insecticide calibration is especially important.

Sampling and decision-making procedures assist producers in determining if a control action is necessary. When control is warranted, the choice of a pest management tactic depends on many factors including the degree of injury from the pest, economic considerations, pest life history, and compatibility of pest management tactics with other production practices. By becoming familiar with sampling, decision making, and the variety of management tactics available, including their proper uses, advantages, and limitations, you will be prepared to make correct pest management decisions when you face insect pest problems.

GENERAL BIBLIOGRAPHY

Fenemore, P. G. 1982. Plant pests and their control. Butterworths, Wellington, New Zealand
- an excellent introduction to pest management with chapters on various management tactics. Insect species mentioned are from New Zealand which limits it usefulness in North America slightly.

Metcalf, R. L., and W. H. Luckmann, eds. 1982. Introduction to insect pest management. 2nd ed. John Wiley and Sons, New York
- chapters on management tactics, pest management principles, and pest management programs for some commodities.

Pimentel, D., ed. 1981. Handbook of pest management in agriculture. Vol. 1-3. CRC Press, Inc., Boca Raton, FL
- a three volume compendium on pest management (insects, weeds, and diseases) with chapters on management tactics and management programs for individual commodities. As is common in multiauthored works, the chapters are quite uneven in their treatment of individual topics.

11 SAMPLING AND DECISION MAKING

Without question, the most important early steps in dealing with an insect pest problem are species identification, assessing the population's damage potential, and deciding on a response appropriate for the situation. A manager's success in performing these steps well can mean the difference between profitability or nonprofitability, environmental conservation or unnecessary intervention, and lasting control or short-term suppression; in other words, success or failure. In this chapter you will learn about some of the most basic elements of population assessment and decision making for insect pest management.

Assessment and decision making is a two-stage process involving information gathering and analysis of this information. Information gathering includes obtaining estimates of pest population density, environmental conditions (e.g., present weather and forecasts), host status (e.g., plant stage and condition), and economic factors (e.g., expected market prices and costs of potential responses). This information is analyzed with a focus on damage potential, and decisions are made using response guidelines from established recommendations or from personally prepared cost/benefit analyses.

PEST POPULATION SAMPLING

After pest identification, the most basic problem is determining the numbers of insects that are present at a given place and time. By gathering this information over an area and for a period of time, projections can be made about the nature of the population and its destructive potential.

Since it would not be feasible to count every single insect in a population, we must make estimates that reflect population density. Estimates are made by **sampling** a population, i.e., taking a representative part of the total and basing our estimate on that part. In addition to direct estimates of pest populations by sampling pests, indirect measures, called population indices, can be used. A **population index** is an estimate of population size based on insect effects or products. A program of sampling to make estimates is called a **survey**, and the persons doing the surveying are usually called **scouts**.

To conduct a survey, both a sampling technique and a sampling program are required. A **sampling technique** is the method used to collect information for a single sample. Swinging an insect sweep net 25 times through an alfalfa canopy to collect a single sample of pea aphids is an example of a technique. A **sampling program** is a method of employing the technique. The sampling program includes such specifications as number of samples to take at a location, when to take the samples, and the spatial pattern to follow in collecting samples.

Common Sampling Techniques

There are many different techniques for sampling insects, each having its own advantages and disadvantages. Most techniques can be grouped into one of six categories:

In situ **counts**. Viewing insects in place (e.g., on the plant or animal) and counting them directly; usually no special equipment is required.

Knockdown. Insects are removed from habitat by jarring, chemicals, or heating, then counted; equipment - **shake** or **beat cloth**, knockdown from plants or plant parts with chemicals or by heating.

Netting. Insects are removed from habitat (plant, air, water) with a net, then counted; equipment - **sweep** or **beating net**, **vacuum net**, **aerial net**, **tow net** (from airplanes and automobiles), **rotary net**, or **aquatic net**.

Trapping. Insects actively move into a device which holds them there for counting; methods may be attractive (baited) or passive (not baited) in capture; equipment -**pitfall**, **window, Malaise, sticky, water pan, pheromone, blacklight**, and **baited sticky traps** (Fig. 11.1).

Extraction from soil. Insects are separated from a quantity of soil by utilizing insect mobility (**Berlese funnel**) and by passive means (sieving, washing, flotation and combinations of these).

Indirect techniques. Quantify the effects of insects or their products; estimates of plant defoliation, ratings of root loss, numbers of cast skins, quantity of frass, and numbers of nests; programs using these techniques produce indices of population size.

Figure 11.1   Example of a sticky trap used for monitoring adult corn rootworm beetle populations in corn.

Sampling Programs

Sampling programs are designed to yield population estimates or indices for estimating damage potentials. Estimates given are of two broad types, absolute and relative. **Absolute estimates** are a measure of the actual insect population according to ground surface area, e.g., number per acre, per hectare, per square meter. Absolute estimates often are costly, and tend to be used most frequently in research on population dynamics and insect-plant relationships. **Relative estimates** are a measure of insect numbers relative to the sampling technique, e.g., numbers per 25 sweeps with a net, numbers caught in a blacklight trap per three nights, and numbers caught in a pitfall trap per day. These estimates do not translate directly into numbers per ground surface area. However, they do allow comparisons of population density from time to time and place to place. Often, they are used to determine pest population trends for pest management decision making once the relationship to absolute numbers is established by research. Relative estimates are usually less expensive to obtain than absolute estimates. Population indices, discussed earlier, are used similarly to relative estimates.

Recommendations for sampling an insect population to obtain an estimate include specific instructions, called **program dimensions**. These dimensions should be followed in making the final population estimate, whether it is absolute or relative. The major program dimensions include: (1) insect stage to sample, (2) sample location, (3) number of samples to take, (4) when to sample, and (5) spatial pattern of sampling. Most often, pest management scouting calls for walking a prescribed route through a growing area, making sure to include samples from high- and low-lying spots and avoiding crop borders.

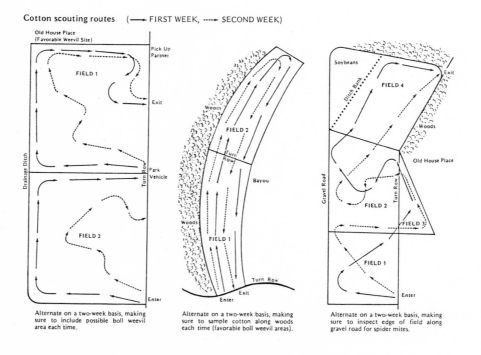

Figure 11.2 Plan for sampling cotton bolls for boll weevil larvae. (Courtesy South Carolina Cooperative Extension Service)

Once numerical information is gathered, these data usually are summarized for each sample date using descriptive statistics. The most widely used statistics for this purpose are the mean, the standard deviation, and the standard error of the mean. The **mean** is simply an arithmetic average based on a sum of the numbers sampled divided by the total number of samples. Although the mean is an important measure of central tendency, it tells us nothing about differences in numbers among different samples. One measure of variation in our samples is the **standard deviation**, a statistic reflecting the average deviation from the mean. When the numbers from many samples can be described by a bell-shaped curve, we would expect that 68 percent of them would fall within one standard deviation on either side of the mean. Another statistic, the **standard error** of the mean, gives us our best estimate of average deviation of numbers in the insect population. It is based on an average deviation of a number of sample means. Standard errors are a measure of our sampling precision (ability to predict the mean within narrow limits). Error can be reduced (estimates made more precise) by increasing the number of samples taken.

After calculating population statistics, means and standard errors can be plotted on graph paper or displayed with computer graphics to detect significant numerical changes and estimate future population trends. Subsequently, this information is used as a basis for management decisions.

Table 11.1   Sequential sampling table for corn rootworm beetles in continuous corn fields, to determine potential rootworm problems in the following year. (Iowa Cooperative Extension Service)

| | NUMBER OF BEETLES FOUND | | |
No. of Plants Sampled	Resample in 7 days	Continue Sampling	Control Needed[1]
10	0-2	3-11	12+
12	0-3	4-12	13+
14	0-4	5-14	15+
16	0-6	7-15	16+
18	0-7	8-17	18+
20	0-8	9-18	19+
24	0-11	12-21	22+
28	0-14	15-23	24+
32	0-16	17-26	27+
36	0-19	20-29	30+
40	0-22	23-31	32+
44	0-24	25-34	35+
48	0-27	28-37	38+
52	0-30	31-40	41+
54	0-31	32-41	42+

[1]To avoid damaging rootworm populations next year, either rotate to a different crop or use a soil insecticide if corn is planted again.

Sequential Sampling

In many instances, management decisions are based on taking a fixed number of samples in an area. An increasingly important alternative to fixed sampling is sequential sampling. **Sequential sampling** is a procedure based on insect dispersion patterns and economic decision levels that uses variable numbers of samples, often fewer samples than in a fixed sampling program. In using a sequential sampling program, the total number of samples to be taken in an area is not known when the sampling is begun. With the aid of a decision table, sampling is started and continued until a decision can be made, i.e., to take curative action or take no action (e.g., Table 11.1). To use the sequential sampling table, you would take a sample and record pest numbers, followed by another sample with its numbers added to those of the previous sample. Accumulations of numbers are made through subsequent samples until the number exceeds a critical value (stop sampling, take action) or falls below a critical value (stop sampling, call the infestation noneconomic, take no action). If no decision can be made within a stated maximum number of samples, you would be directed to stop sampling and return to the location in a few days to sample again. Although intermediate populations may still require considerable sampling, an average savings in sampling costs of 50 percent often is achieved by using sequential sampling programs.

PREDICTING BIOLOGICAL EVENTS

One of the most important elements of monitoring insect populations for decision making is knowing when to begin the sampling program. Some sampling programs are initiated on a given calendar date. Another approach to proper timing is to consider temperature and time by calculating values called degree days, thereby being able to predict important biological events in the insect life cycle. For example, after accumulating 300 degree days from the first significant black cutworm moth flight corn producers would be advised to scout fields for damaging larval populations.

Degree Days Defined

Two of the most important factors in the growth of any organism are time and temperature. Although temperature is important in growth of warm-blooded animals, its significance is not apparent because body temperature is constant. Conversely, the relationship between temperature and growth is dramatic in cold-blooded forms. Organisms such as insects and plants cannot maintain a constant temperature, and body temperatures will vary with the temperature of the environment. Within limits, these organisms grow faster with warmer temperatures.

Consequently, by quantifying this relationship we can use measures of environmental temperature to predict growth. However, time and temperature must be combined in some way to make these growth estimates. A combination of growth and time is called **physiological time**, and degree days can be used as a means to predict physiological time.

Degree days represent the accumulation of heat units above some minimum temperature for a 24-hour period. Below the minimum, no development takes place, but above it, accumulations are made toward development. Ten degrees above the minimum for 5 days represents 50 degree days (10 x 5), as does 2 degrees above the minimum for 25 days (2 x 25). In either instance an insect would have grown the same amount even though the real time was quite different.

Calculating Degree Days

To calculate degree days we must recognize that growth only occurs within a range of temperatures. The minimum temperature below which no growth occurs is called the **minimum developmental threshold**. Insect growth will increase with higher temperatures up to a maximum temperature called the **maximum developmental threshold**. These threshold temperatures are determined experimentally and are listed in publications that recommend degree days for predicting growth.

The simplest and most common approach to degree-day calculation utilizes a technique called the rectangle, or historical, method. With this method a simple average is calculated from the daily maximum and minimum temperature, and the developmental threshold is subtracted from this average:

degree days = ([daily max. + daily min. temperature]/2) - minimum threshold

If minimum temperature < minimum threshold, set minimum temperature = minimum threshold.
If maximum temperature > maximum threshold, set maximum temperature = maximum threshold.

EXAMPLES

Black Cutworm Development
Developmental thresholds: minimum = 50°F and no maximum
Daily minimum and maximum: 62°F and 90°F
Calculation: ([62 + 90]/2) - 50 = 26 degree days

Corn Development
Developmental thresholds: minimum = 50°F and maximum = 86°F
Daily minimum and maximum: 62°F and 90°F
Calculation: ([62 + 86]/2) - 50 = 24 degree days

Seedcorn Maggot Development
Developmental thresholds: minimum = 39°F and maximum = 84°F
Daily minimum and maximum: 34°F and 58°F
Calculation: ([39 + 5]/2 - 39 = 9.5 degree days

Degree days are totaled over a period of days to determine when an insect has reached a certain stage. Accumulations called **thermal constants** are stated for each stage of insect development. Thermal constants differ for different stages of a species and are different between species. Dates to start accumulations also differ between species, but a common starting time is when the daily maximum exceeds the minimum threshold or when migratory insects are detected in an area.

The major use of degree days in insect pest management is in timing the scouting of pest species. By accumulating degree days we can eliminate unnecessary scouting, avoid overlooking injurious pest populations, and make better management decisions.

MAKING MANAGEMENT DECISIONS

After accumulating information on insect numbers in appropriate stages, the next step is to compare the numbers with those of representing activity guidelines. Such activity guidelines usually are expressed as number of insects per area, number per plant or animal unit, or number per sampling unit.

The most widely used activity guidelines are presented as economic thresholds. The **economic threshold** (ET) is the number of insects that calls for corrective action. It is based on the **economic-injury level** (EIL), the number of insects that would cause damage equal to cost of dealing with those numbers, i.e., the break-even level. The ET is usually set lower than the EIL, at a point where densities, if left unchecked, will likely exceed the EIL.

Both of these levels are established from an analysis of economic damage. **Economic damage** has been defined as the amount of injury that will justify the cost of artificial control measures. Economic damage begins to occur when money required for alleviating insect injury is equal to money lost in damage from a pest population. Economic damage can be expressed in terms of the commodity damaged using the gain threshold. The **gain threshold** is represented as:

$$\text{gain threshold} = \text{management costs/market value}$$

for example,

$$\text{gain threshold} = (\$9/\text{acre})/\$3/\text{bushel} = 3 \text{ bushels/acre}$$

In other words, at least 3 bushels per acre would need to be saved with with an insecticide application or other activity for control actions to be profitable. Therefore, the gain threshold is our worksheet standard; our margin for determining benefits of management. The EIL is the number of insects that will cause damage equal to the gain threshold. By calculating the gain threshold and knowing the loss per insect, we can calculate the EIL as follows:

$$\text{EIL} = \text{gain threshold/loss per insect}$$

Obtaining the loss per insect is the most difficult aspect of the EIL calculation. Estimates of losses are obtained from field observation or experimentation with various sized insect populations. Such information, as well as precalculated EILs, can often be acquired from publications of the Cooperative Extension Service in each state. After calculation of the EIL, the ET can be set at a level conservatively below the EIL, for instance at 75% of the EIL. Subsequently, action should be taken against an insect species when numerical estimates of actual population size (or an index of it) equal the ET, and the trend in population growth indicates that numbers will exceed the EIL. Such an approach helps avoid significant monetary losses while reducing the chances of unnecessary insecticide applications.

OBJECTIVES

1) Understand the basic principles of insect population sampling, be able to make population estimates, and develop skills in analyzing numerical trends.

2) Obtain knowledge of insect development and its relationship to environmental temperature and apply this knowledge to estimate insect growth stages.

3) Learn about important management-decision rules and gain ability to decide on appropriate responses to pest problems.

4) Additional objectives (at instructor's discretion):

LIST OF DISPLAYS AND EXERCISES

1) Sampling Equipment. Observe equipment commonly used to sample insects and become familiar with techniques and learn how to use a sweep net. Instructions on page 225.

2) Sampling a Population and Describing Trends. Learn to estimate population size over time and predict future trends. Instructions on page 225.

3) Decision-making with Sequential Sampling. Learn to sample a population until a decision can be made. Instructions on page 225.

4) Predicting Events in an Insect Population. Be able to calculate degree days and estimate the time a sampling program should begin. Instructions start on page 226.

5) Calculating Economic Injury Levels and Establishing Thresholds. Learn to calculate economic-injury levels and establish economic thresholds. Instructions on page 226.

OPTIONAL DISPLAYS AND EXERCISES

1) European Corn Borer Management Model. Making pest management decisions by using a management model for European corn borer personal computer. Instructions start on page 227.

2) Examine displayed examples of sampling programs and economic-injury levels for insect pests in your area.

3) Additional exercises (at instructor's discretion):

TERMS

absolute estimate
aerial net
Berlese funnel
blacklight trap
degree day
developmental
   threshold
economic-injury level
economic threshold
extraction from soil
gain threshold

indirect techniques
in situ count
knockdown
Malaise trap
mean
netting
pest survey
pheromone trap
physiological time
pitfall trap

population density
population index
program dimensions
relative estimate
sampling
sampling program
sampling technique
scout
sequential sampling
shake or beat cloth

standard deviation
standard error
sticky trap
sweep net
thermal constant
tow net
trapping
vacuum net
water pan trap
window trap

DISCUSSION AND STUDY QUESTIONS

1) List some techniques that might be used in sampling programs to obtain absolute estimates. Which techniques could be used for relative estimates? Why aren't absolute estimates always the best type to use?

2) Look up and list formulas for calculating the mean, standard deviation, and standard error of the mean. What are the uses of these statistics?

3) Discuss how the physiological time concept is important for management of insect pests in your locality and specialty. How useful would the degree day concept be in tropical environments?

4) Describe how the economic-injury level changes with regard to changing crop values, management costs, and host response to injury.

5) Discuss ways that personal computers might be useful to producers in making pest management decisions.

BIBLIOGRAPHY

Pedigo, L. P. 1989. Entomology and pest management. Macmillan Pub. Co., New York, NY
   - Chapter 5. Insect ecology, Chapter 6. Surveillance and sampling, and Chapter 7. Economic decision levels for pest populations.

Bechinski, E. J., G. D. Buntin, L. P. Pedigo, and H. G. Thorvilson. 1983. Sequential count and decision plans for sampling green cloverworm (Lepidoptera: Noctuidae) larvae in soybean. J. Econ. Entomol. 76: 906-812
   - Gives an example of a sequential sampling table.

Higley, L. G., L. P. Pedigo, and K. R. Ostlie. 1986. DEGDAY: a program for calculating degree-days, and assumptions behind the degree-day approach. Environmental Entomology. 15:999-1016

- Presents several methods of degree-day calculations and gives rationale behind the approach. A program for use on personal computers is listed.

Kogan, M., and D. C. Herzog, eds. 1980. Sampling methods in soybean entomology. Spring er-Verlag, New York, NY

- Although emphasizing insects of soybeans, has good sections on principles of sampling and most major sampling techniques.

Pedigo, L. P., S. H. Hutchins, and L. G. Higley. 1986. Economic injury levels in theory and practice. Annu. Rev. Entomol. 31:341-368

- A review of the topic of decision guidelines in pest management with many examples.

Raupp, M. J., J. A. Davidson, C. S. Koehler, C. S. Sadof, and K. Reichelderfer. 1987. Decision-making considerations for aesthetic damage caused by pests. Bull. Entomol. Soc. Am. 34:27-32

- an excellent article that explores the dilemma of when to control aesthetic pests. The authors describe a technique for addressing aesthetic pests in the context of an economic injury level.

Southwood, T. R. E. 1978. Ecological methods with special reference to the study of insect populations. 2nd ed. Chapman and Hall, New York, NY

- A major reference on sampling techniques and sampling programs.

Stern, V. M., R. F. Smith, R. van den Bosch, and K. S. Hagen. 1959. The integrated control concept. Hilgardia. 22:81-101

- A classic paper that introduced the economic-injury level concept.

DISPLAYS AND EXERCISES

Exercise I. Sampling Equipment

Examine the sampling equipment on display. Identify the major technique category to which each piece belongs. Try swinging the sweep net in a figure-8 pattern as you walk along, as if sampling in a plant canopy. Have your instructor check your skill.

Exercise II. Sampling a Population and Describing Trends

Attend a laboratory station having containers of brown beans and white beans. Brown beans represent pests in the crop. Start with the container marked date 1 and take five samples (a sample is obtained with the scoop). Record number of pests in each sample on columned paper. Use the same sampling method for the remaining dates (dates 2 through 5) and record the numbers. Calculate the mean for each date and plot the means on graph paper. What is the trend in the population? If this population trend continues, what would be your prediction of population density on date 6?

Exercise III. Decision-making with Sequential Sampling

Go to a laboratory station having a container marked sequential sampling. Take samples of pests (brown beans) with the scoop and accumulate numbers with each succeeding sample. Beginning with the third sample, compare accumulations with the numbers in the following sequential sampling table. Stop sampling when a management decision can be made. Based on your sampling, what would be your recommendation? If this program had been used rather than a fixed program of 15 samples, would savings have resulted? If so, how much?

Sample Number	Cumulative Pests in Sample		Low Level		High Level	
1	_____		---		---	
2	_____		---		---	
3	_____		22		30	
4	_____		31		38	
5	_____		40		47	
6	_____		48		56	
7	_____	Take	57	Indecision	64	
8	_____	No	66	Zone	73	Spray
9	_____	Action	74		81	
10	_____		83		89	
11	_____		92		98	
12	_____		100		107	
13	_____		109		115	
14	_____		118		124	
15	_____		126		133	

Exercise IV. Predicting Events in an Insect Population

Assume that you are responsible for sampling an insect pest for assessment of pest status when individuals in a population are in the third larval stage. The third larval stage occurs at a thermal constant of 117 degree days. The developmental threshold for this insect is 52°F, and degree day accumulations are begun after the first adult is collected in a blacklight trap. In this instance the first adult was collected on the night of May 21. Based on the following records of daily maximum and minimum temperatures, when would you begin sampling to determine pest status?

Date	Max. Temp.	Min. Temp.	Degree Days	Accumulative Degree Days
May 20	74	48		
May 21	77	52	_____	_____
May 22	67	49	_____	_____
May 23	51	37	_____	_____
May 24	66	48	_____	_____
May 25	70	51	_____	_____
May 26	78	61	_____	_____
May 27	79	62	_____	_____
May 28	80	63	_____	_____
May 29	81	60	_____	_____
May 30	79	59	_____	_____
June 1	83	65	_____	_____
June 2	82	60	_____	_____
June 3	84	61	_____	_____

Exercise V.

Calculating Economic Injury Levels and Establishing Thresholds

Suppose that you also are required to develop economic thresholds for the pest population you are monitoring. If the economic threshold is set at 80% of the economic-injury level, what would be its value for the crop under the following circumstances:

Expected crop market value $30/bushel
Estimated management costs $12/acre
7 insects/plant causes 2 bushels/acre loss

EIL = _____   ET = _____

What if market value unexpectedly dropped to $25/bushel?

EIL = _____   ET = _____

What if management costs increased by $3/acre?

EIL = _____   ET = _____

Exercise VI. European Corn Borer Management Model Exercise

Recently, computer-based management guides have been advocated as a desirable approach for incorporating numerous factors into pest management decisions. One example of such a management tool is the European corn borer management software developed by Kansas State University. (This software is available from the Computer Systems Office, 211 Umberger Hall, Kansas State University, Manhattan, KS 66506.)

The European corn borer (ECB), *Ostrinia nubilalis*, overwinters as large larvae in plant debris on the soil surface. Adult corn borers emerge in early summer and produce first generation ECB larvae shortly after emergence. These larvae may produce sufficient injury to corn that control action should be taken. The threshold for first generation ECB is based on the number of plants infested and a yield loss of 5% for one corn borer per plant. The second generation of ECB usually occurs in mid to late summer and also can produce economic losses. Unfortunately, determining when to sample and economic losses associated with a given ECB population is quite difficult. The software you will use in this exercise will allow you to make sampling and management decisions for second generation ECB. This software incorporates information on life history, natural control, ECB development (by degree days), sampling, and economic injury levels. Thus, it integrates many of the topics you've studied. With greater knowledge and refinement of pest management systems, programs such as this are likely to become more available and more appropriate for making pest management decisions.

How to Get Started

To start, at the system prompt enter the name of the program which is "ECBEXTN" and hit return. The first question the program will ask is if you are using a color monitor. After a few screens of copyright information, etc. the program will prompt you for the date. When entering months for this and subsequent date entries, remember to use a zero before a single digit month. Thus, enter April 30, 1988 as 04/30/88. If you have any questions about what the program is requesting or what a certain input means, just enter "H" (for Help) and the program will provide information.

Using the Program

Two models are available in the program: a phenology model, which predicts when sampling should be conducted to estimate second generation ECB populations; and a management model, which determines whether or not control action is warranted. You will run both models (the management model uses information from the phenology model so you need to run the phenology model first). You are strongly encouraged to use the help option periodically, as the program can provide considerable information on the meaning and rational of inputs and results.

Assume you are a farmer in northwestern Kansas with potential ECB problems. You are going to use the program to evaluate those problems. Sample data for this exercise are taken from actual reports from Kansas. First, you will use the phenology portion of the program to predict ECB oviposition (so we know when to start sampling).

Enter "P" to begin the phenology model. The program will prompt for a field name, you can call it whatever you like. Next the program will want a county name. For this example, you are farming in Brown county Kansas, so enter "Brown". Now you need to enter information on the sample date, age and number of first generation ECB in our field. Enter 06/18 for the sample date. From this sampling you found 2 1st instars, 28 2nd instars, 15 3rd instars, 16 4th instars, 3 5th instars, and 0 pupae. Notice that the program issues a warning that you should resample in a few days, but you can just take your chances with the first sample. Next the program wants temperature information so it can predict (using degree days) ECB development. Just use the 30-year average. At this point enter "R", the program will run, and you can note the recommended sampling dates.

Now you can run the management model. Type "M". Assume you took your sample during the recommended sampling interval, specifically on 07/16. You noted 15 egg masses on 100 plants. The corn is just at silking, and the first day at silking was probably the same day you sampled, 07/16. Your corn variety has a relative maturity of 110 days. You estimate that there were 20 eggs per egg mass. The last information you have to enter is the % survivorship of ECB and the % ECB control. For this, use the suggested values in the program (20% and 67%, respectively). The next set of inputs required are economic variables. For corn yield, use 140 bushel/acre with an estimated price per bushel of $1.48. You now need to put in your insecticide costs. Among the labeled compounds available are:

Dyfonate 20G   ca. $ 9.92/acre
Furadan 4F   ca. $12.37/acre
Pydrin 2.4EC   ca. $ 6.29/acre

For this first run use Dyfonate at $9.92/acre. The program will also prompt for application costs. Aerial application is most common, and you can estimate this cost at $5.00/acre. Finally, you will need a single insecticide application. That completes the data entry and you can run the management model by entering "R".

Answer the following questions and take some time to experiment with the inputs to see how the program can be used to examine different alternatives such as insecticide costs, corn price or yield, different levels of ECB infestations, etc.

Questions

1. What is the recommended sampling interval (dates) based on the phenology model?

2. What is the cost/benefit ratio from the management model? What is the control recommendation?

3. Change the insecticide cost to $6.29/acre (Pydrin instead of Dyfonate) and run the management model again. Now what is the cost/benefit ratio and control recommendation?

4. Change the biological parameters so you have 20 egg mass per 100 plants (rather than 15 per 100). Now what is the cost/benefit ratio and control recommendation (using Pydrin)?

5. What is your impression of this model? Do you believe it would be of practical benefit to growers? What drawbacks to you see to its use?

12 INSECTICIDAL MANAGEMENT

Insecticide application is the most powerful and most popular tactic used in managing insect pests. Insecticides have a number of advantages over other pest management tools; they are effective and reliable, rapid in their action, easy to apply, flexible in meeting changing agronomic and ecological conditions, and, usually, economical. Also, insecticides are the only pest management tool that is feasible for an emergency or **rescue treatment** (i.e., when insect pest populations approach or surpass the economic threshold).

Unfortunately, insecticides are not a panacea for all pest management problems. Insecticide use has several inherent disadvantages that must be recognized. For example, repetitive applications of insecticides can lead to decreasing insecticide effectiveness due to (1) the genetic acquisition of **resistance** to insecticides in repetitively-exposed populations of insects; (2) the **resurgence** of pest populations following insecticide applications that interfere with natural control agents (predators and parasites); and (3) the **replacement** of the suppressed target pests with other, previously unimportant, secondary pests.

Insecticides also may have adverse effects on nontarget species including honeybees, other pollinators, and wildlife species. The accumulation of insecticide residues in the environment, and especially in groundwater supplies, is another concern and possible risk associated with insecticide use. Finally, insecticides may present direct risks to the user. Many insecticides are very toxic to human beings and can injure or kill if handled, applied, or disposed of improperly. Consequently, the federal and state governments in the US have instituted a number of laws regulating the use and classification of pesticides including insecticides, the certification of pesticide applicators, the transportation and disposal of pesticides, and the pesticide residues permitted on foods. A summary of the pertinent laws relating to pesticides and their use is listed in Appendix 12.1 at the end of this chapter.

INSECTICIDES

Insecticide Routes of Entry

Insecticides can be conveniently grouped according to their route of entry into the insect. Most insecticides are **contact poisons**. Contact insecticides enter the body when the insect walks or crawls over a treated surface. The insecticide must be absorbed through the integument of the insect and therefore must be a relatively fat-soluble, or **lipophilic**, compound. Contact poisons are frequently applied as liquid sprays or dusts to the insect pest's hosts or to other surfaces that the pest is likely to contact. Contact poisons generally have **residual activity**, or remain toxic for some time following application.

Stomach poisons enter the insect body through the gut and are fatal only after they are eaten. They are therefore commonly the active ingredient in **bait** formulations and in **systemic insecticides**. Systemics are insecticides that are taken up and translocated within plant or animal hosts of insect pests. They enter the insect pest via the gut when the insect feeds upon

218

a treated host. Systemic insecticides are typically water-soluble, or **hydrophilic**, so that they can move freely within the hosts' vascular systems. Ideally, systemic insecticides should not be toxic to the treated hosts.

Fumigants enter the insect through the spiracles and tracheal system. Fumigants are volatile insecticides that become gases at temperatures above $5^{\circ}C$. They are applied to enclosures, such as grain bins, and to the soil. Fumigants have little or no residual activity; as soon as the gas diffuses from the treated area its effectiveness is lost.

Insecticide Mode of Action

Insecticides may also be grouped according to their **mode of toxic action**, or how they work in causing harm or death once they have entered the insect body. In terms of mode of action, insecticides may be classified as **selective** or **nonselective**. Selective insecticides exert an effective toxic action against a limited group of organisms. An example of a selective insecticide is the toxin produced by *Bacillus thuringiensis*, which is detrimental primarily to lepidopteran larvae. This selectivity is termed "physiological selectivity". Nonselective insecticides are toxic to a variety of organisms and are therefore nondiscriminating in their action. Most of the organophosphorous insecticides are nonspecific, or **broad-spectrum**, and may harm or kill a great variety of insects and other organisms. In a well-designed pest management program, it is best to choose the most selective insecticide available to prevent disruption of populations of natural enemies or other nontarget species. Unfortunately, the most effective and widely-used insecticides are relatively nonselective. Selectivity, therefore, must come from proper application, i.e., applying insecticides in amounts, at times, and in places that limit their contact by natural enemies and other nontarget species but enhance their contact by the target pest. This type of selectivity is termed "ecological selectivity".

The majority of today's insecticides are **neurotoxic** and relatively nonselective in their mode of action. The nervous system is a susceptible and vulnerable target, and poisoning it is a rapid and dependable means of disrupting a pest's life processes. Neurotoxic insecticides act primarily by: (1) interfering with impulse transmission along the axons of nerve cells (e.g., DDT and analogs, pyrethroids) or (2) interfering with impulse transmission at the junctions, or **synapses**, between nerve cells (e.g., organophosphates, carbamates, nicotine). The **muscle poisons**, which include some botanical insecticides (e.g., ryania, sabadilla) act similarly: these materials disrupt the excitable membrane, or sarcolemma, that surrounds muscle cells.

Essential metabolic pathways are common targets for many insecticides. **Metabolic inhibitors** function by blocking or uncoupling necessary sequences of metabolism. Most metabolic inhibitors are nonselective since they disrupt pathways common to all living organisms. Examples of metabolic inhibitors include: methyl bromide, chloropicrin, and fluoroacetate which disrupt the Kreb's cycle; fluoride which impairs glycolysis; and rotenone and cyanide which interfere with the electron transport chain. Some metabolic inhibitors do exhibit a substantial degree of selectivity. For instance, materials such as diflubenzuron can impair pathways, such as those involved in chitin synthesis, that are fairly unique to insects. Insect hormonal systems are also suitable targets for insecticides. The mode of action of the insect growth-regulating compounds such as methoprene, a juvenile hormone mimic, is interference with these systems. Hormone mimics, analogs, or antagonists offer a very high degree of selectivity but have not yet been heavily exploited commercially.

Physical toxicants function by interfering with a life process, such as respiration, by physical, rather than chemical, means. The selectivity of physical toxicants depends primarily on the method of application and the placement of the material. Common physical toxicants include oils, used to clog the spiracles of insects, and abrasive dusts and powders, used to disrupt cuticular structure and induce subsequent dehydration.′ Examples of physical poisons include oils applied to trees for scale insect control or to water for management of mosquito larvae, and silica gel and boric acid used to control cockroaches in areas where insecticide residues are not desired.

Insecticide Formulation

The **active ingredients** (a.i.) of insecticides are the chemicals that exert the toxic actions that harm or kill insects. Active ingredients rarely can be used alone and generally are mixed with auxiliary materials to make them convenient and safe to handle, and accurate and easy to apply. The mixture of the active ingredients and auxiliary materials is called the **insecticide formulation**. Most auxiliary substances are **inert**, or have no direct effect on pests, serving strictly as carrier materials for the active ingredients. Other auxiliary materials, however, have important properties and functions. **Synergists**, for example, increase the toxicity of the active ingredients but are, themselves, relatively nontoxic. Various **surfactants** improve emulsifying, wetting, and spreading properties of the insecticide formulation, whereas **stickers** enable active ingredients to be retained on treated surfaces for longer periods.

There are many types of insecticide formulations available. A summary of the most common formulations and their properties is presented in Appendix 12.2. In choosing the formulation that is best for a particular use the following should by considered: (1) the plant, animal, or surface to be protected (some formulations may be phytotoxic, absorbed by the animal, or may pit and mar surfaces); (2) the application machinery available and best suited for the job (some formulations require constant agitation or specialized equipment); (3) the hazard of drift or run-off (proximity to sensitive areas and the likelihood of wind or rain is important); (4) the safety to applicator and other humans and pets likely to be exposed; (5) the habits or growth patterns of the pest; (6) the cost; and (7) the type of environment in which application must be made (agricultural, urban, aquatic, etc.). The label on an insecticide product will indicate what type of formulation the package contains. The same insecticide often is available in more than one formulation.

USING INSECTICIDES

Using insecticides effectively begins with a proper strategy. With agricultural pests, insecticide use often is a curative tactic that should be applied only after pest status has been assessed. Assessment involves accurate identification of the pest species present and an estimation of its abundance. Moreover, information on other potential pests, as well as natural enemies, in the agroecosystem should be part of the overall analysis. If the assessment shows that one or more pests will likely exceed the economic injury level, and no other tactic is practical, an appropriate insecticide should be chosen.

′ Choosing the most appropriate insecticide depends on several factors including effectiveness, cost, formulations available, equipment required, hazards of use, and most importantly, whether or not the product is labeled for the insect and crop or circumstance for which application is intended. The Cooperative Extension Service is an excellent source for recommendations regarding the best product(s) to employ in a given situation.

AUGUST 1987

RESTRICTED USE PESTICIDE
⑨ Due to Acute Oral and Dermal Toxicity and Bird Toxicity

For retail sale to and use only by Certified Applicators or persons under their direct supervision.

American Cyanamid Company endorses Certification to promote the responsible
use of pesticides to insure the protection of man and the environment.

LOW ODOR
① **THIMET** ®
20-G ②
soil and systemic insecticide
SPECIMEN

③ Active Ingredient:
④ Phorate (0,0-diethyl S-[(ethylthio) methyl] phosphorodithioate). 20.0%
Inert Ingredients 80.0%
Total 100.0%
EPA Reg. No. 241-257-AA   EPA Est. No.241-MO-1
⑤   **KEEP OUT OF REACH OF CHILDREN**   ⑤

⑥ **DANGER!** **POISON**

¡PELIGRO!

PRECAUCION AL USUARIO: Si usted no lee inglés, no use este producto hasta que la etiqueta le haya sido explicada ampliamente.

FIRST AID

If swallowed, drink one or two glasses of water and induce vomiting by touching back of throat with finger. Do not induce
vomiting or give anything by mouth to an unconscious person. Avoid alcohol. Get medical attention.

If inhaled, remove to fresh air. If not breathing, give artificial respiration, preferably mouth-to-mouth. If breathing is difficult, give
oxygen. Get medical attention.   ⑦

If on skin, wash thoroughly with soap and water. Remove contaminated clothing and shoes. Wash clothing and decontaminate
shoes before reuse.

If in eyes, immediately flush eyes with plenty of water. Get medical attention.

NOTE TO PHYSICIANS: Warning symptoms include weakness, headache, tightness in chest, blurred vision, nonreactive pin-
point pupils, salivation, sweating, nausea, vomiting, diarrhea and abdominal cramps. Give atropine intramuscularly or intra-
venously, depending on severity of poisoning, 2 to 4 milligrams every 10 minutes until fully atropinized as shown by dilated
pupils, dry flushed skin and tachycardia. Twenty to thirty milligrams, or more, may be required during the first 24 hours. Never
give opiates or phenothiazine tranquilizers. Clear chest by postural drainage. Artificial respiration or oxygen administration may   ⑦
be necessary. Observe patient continuously for at least 48 hours. Allow no further exposure to any cholinesterase inhibitor until
cholinesterase regeneration has taken place as determined by blood test.

Pralidoxime chloride (2-PAM; PROTOPAM chloride) may be effective as an adjunct to atropine. Use according to label directions.

Antidote: Atropine is an antidote
See Back Panel for Additional Precautionary Statements.

Net Weight: 50 lbs.
   22.68 kg. ⑧

24638-20   D50

Figure 12.1   Sample insecticide label. (Courtesy American Cyanamid Company)

PRECAUTIONARY STATEMENTS
HAZARDS TO HUMANS AND DOMESTIC ANIMALS
DANGER!

Fatal if swallowed, inhaled, or absorbed through the skin. Do not breathe dust. Do not get in eyes, on skin, or on clothing.

Wear freshly laundered, long-sleeved work clothing daily. While transferring from package to equipment, wear a clean cap, gloves (rubber or cotton) and goggles. If cotton gloves are used, they must be laundered or discarded after each day's use. Rubber gloves should be washed with soap and water after each use. Do not wear the same gloves for other work. Destroy and replace gloves frequently.

In case of contact, immediately remove contaminated clothing and wash skin thoroughly with soap and water. Launder clothing and decontaminate shoes before reuse. Wash thoroughly with soap and water before eating or smoking. Bathe at the end of the work day and change clothing.

Do Not Breathe Dust

Wear a face mask or other respiratory equipment while emptying bags of product into a hopper. While emptying bags into equipment, pour downwind and allow as little free fall as possible. Do not pour at face level and do not allow dust to reach the breathing zone.

Do Not Contaminate Food or Feed Products

Once a bag has been opened, use it completely. Make sure the hoppers are emptied while still in the field. Refer to STORAGE AND DISPOSAL statement for further instructions.

Keep All Unprotected Persons Out of Operating Areas.

Do not apply this product in such a manner as to directly or through drift expose workers or other persons.

Keep Out of Reach of Domestic Animals

Not For Use or Storage In or Around the Home

RE-ENTRY STATEMENT

Do not enter treated areas without protective clothing until treatments have been completed.

Because certain states may require more restrictive re-entry intervals for various crops treated with this product, consult your State Department of Agriculture for further information.

Written or oral warnings must be given to workers who are expected to be in a treated area or in an area about to be treated with this product. When oral warnings are given, warnings shall be given in a language customarily understood by workers. Oral warnings must be given if there is reason to believe that written warnings cannot be understood by workers. Written warnings must include the following information: DANGER. Area treated with THIMET on (date of application). Do not enter without protective clothing.

Fields may be re-entered on day of treatment with soil applications of THIMET 20-G. Do not enter corn fields within 7 days after plants have been treated with foliar applications.

ENVIRONMENTAL HAZARDS

Product is toxic to fish, shrimp, crab, birds and other wildlife. Birds and other wildlife in treated areas may be killed. Keep out of lakes, streams, ponds, tidal marshes and estuaries. Do not apply where runoff is likely to occur. Do not apply when weather conditions favor drift from areas treated. Do not contaminate water by cleaning of equipment or disposal of wastes. Shrimp and crab may be killed at application rate recommended on this label. Do not apply where these are important resources. Apply this product only as specified on this label.

This pesticide is toxic to bees exposed to direct application. Applications should be timed to coincide with periods of minimum bee activity, usually between late evening and early morning.

STORAGE AND DISPOSAL
⑫

Storage
Store pesticide products in a secure locked area where children, unauthorized persons and animals cannot enter. Do not store in the same area with food or feed. Do not store opened bags.
Prohibitions
Do not contaminate water, food or feed by storage or disposal. Open dumping is prohibited. Cover or incorporate spills.
Pesticide Disposal
Pesticide wastes are acutely hazardous. Improper disposal of excess pesticide, spray mixture, or rinsate is a violation of Federal Law. If these wastes cannot be disposed of by use according to label instructions, contact your state pesticide or Environmental Control Agency, or the hazardous waste representative at the nearest EPA Regional Office for guidance.
Container Disposal
Completely empty bag into application equipment. Then dispose of empty bag in a sanitary landfill or by incineration, or, if allowed by state and local authorities, by burning. If burned, stay out of smoke.
General
Consult federal, state or local disposal authorities for approved alternative procedures such as limited open burning.

CALL A PHYSICIAN AT ONCE
IN ALL CASES OF SUSPECTED POISONING.

 DIRECTIONS FOR USE

⑭ It is a violation of Federal Law to use this product in a manner inconsistent with its labeling.

This label must be in the possession of the user at the time of pesticide application.

BEFORE USING, READ PRECAUTIONARY STATEMENTS.

THIMET should be applied with a granular pesticide applicator properly calibrated to assure accurate placement and proper dosage. Cover granules that may be exposed on the ends of the treated rows and turns and loading areas by deep discing immediately after treating fields.

Figure 12.1 Sample insecticide label, continued. (Courtesy American Cyanamid Company)

CROPS

⑬

Crop	Pests Controlled	Rate of THIMET 20-G	Application	Remarks
BEANS At Planting	Mexican bean beetles	4.9–9.4 ozs. per 1,000 ft. of row for any row spacing (minimum 30-inch spacing).	Drill granules to the side of the seed.	Do not place THIMET granules in direct contact with seed. Do not feed the foliage of treated beans within 60 days of treatment.
	Leafhoppers Aphids Lygus bugs Thrips Mites	4.5-7.0 ozs. per 1,000 ft. of row for any row of spacing (minimum 30-inch spacing)		
	Seedcorn maggots		Apply granules in a band over the row.	
FIELD CORN AND SWEET CORN (excluding popcorn) At Planting	Corn rootworms Wireworms White grubs Seedcorn maggots Seedcorn beetles Flea beetles Mites	6 ozs. per 1,000 ft. of row for any row spacing (minimum 30-inch row spacing)	Place granules in a 7-inch band over the row, in front of or behind the press wheel and lightly incorporate.	Do not place THIMET granules in direct contact with seed. One additional application can be applied at cultivation OR over corn later in the season.
FIELD CORN AND SWEET CORN (excluding popcorn) At Cultivation	Corn rootworms		Apply granules to base of plants or over the top of plants just ahead of cultivation shovels so as to cover granules with soil.	Do not make any applications of THIMET after cultivation treatment.
FIELD CORN (excluding sweet and popcorn) At Cultivation	Chinch bug nymphs (Colorado, Kansas, and Nebraska)		Apply granules to base of plants or over the top of plants just ahead of cultivation shovels so as to cover granules with soil. Apply as soon as chinch bug nymphs are observed at base of plants. Granules must be placed at the infested region of the plants and adequately covered with soil.	Later infestations of chinch bugs may not be controlled. Do not make any applications of THIMET after cultivation treatment.
FIELD CORN (excluding sweet and popcorn) Over the Plant	European corn borer (1st brood)	5.0 lbs. per acre	Apply granules into the whorl of the plant prior to tassle emergence.	Do not graze or cut for forage within 30 days of treatment. Do not make more than one application over the plant. Do not apply under prolonged drought conditions. Consult your state experiment or state extension service for proper timing of application.
	Corn leaf aphids Mites		Broadcast the granules evenly over the top of plants by air or ground equipment.	
COTTON At Planting	Early season control of: Thrips Mites Leaf miners	3.0 ozs. per 1,000 ft. of row for any row spacing (minimum 36-inch spacing)	Distribute the granules evenly in the furrow at planting time. Soil temperature, moisture and tilth should be favorable for good germination and emergence.	Suggested for areas where a limited period of protection against early season pests is necessary for good cotton growth. Use 1.5 lbs. THIMET 20-G per acre for hill-dropped cotton.
	Mites Thrips Aphids Black cutworms Leafhoppers White flies Leaf miners	6.0-9.0 ozs. per 1,000 ft. of row for any row spacing (minimum 36-inch spacing)	Distribute the granules evenly in the furrow at planting time. Soil temperature, moisture and tilth should be favorable for good germination and emergence. Under adverse conditions plant stand may be affected.	Recommended for areas requiring an extended period of mite control as well as early season protection against other cotton pests. The higher rate is suggested for heavier soils.
COTTON Side-Dressing (Irrigated Cotton Only)	Mites from spring to early summer	9.0-12.0 ozs. per 1,000 ft. of row for any row spacing (minimum 36-inch spacing)	Incorporate granules into the soil as a side-dress application deep enough to avoid disturbance by future cultivations.	Irrigate as soon as possible after treatment. Do not apply later than 60 days before harvest.
PEANUTS At Planting	Thrips Leafhoppers	5.5 ozs. per 1,000 ft. of row for any row spacing (minimum 24-inch spacing)	Distribute the granules evenly in the furrow at planting time. For best results, plant peanuts when soil temperatures are favorable for establishing good peanut stands.	Do not graze or feed treated hay or forage.

3

Figure 12.1   Sample insecticide label, continued. (Courtesy American Cyanamid Company)

CROPS ⑬

Crop	Pests Controlled	Rate of THIMET 20-G	Application	Remarks
PEANUTS At Pegging	Southern corn rootworms and Leafhoppers	11.0 ozs. per 1,000 ft. of row for any row spacing (minimum 24-inch spacing)	Distribute the granules as a band over the fruiting zone at pegging time. Work into the top few inches of soil immediately.	Do not graze or feed treated hay or forage.
POTATOES At Planting	Aphids Leafhoppers Leaf miners Psyllids Wireworms Flea beetle larvae and reduction of flea beetle adults	Light or Sandy Soils: 11.3 ozs. per 1,000 ft. of row for any row spacing (minimum 32-inch spacing) Heavy or Clay Soils: 17.3 ozs. per 1,000 ft. of row for any row spacing (minimum 32-inch spacing)	Distribute the granules evenly in the furrow or band on each side of the row.	Wait 90 days after treatment before harvesting potatoes. Do not use for Colorado potato beetle control in the Northeast United States.
	Colorado potato beetle (early season control)	Heavy or Clay Soils: 17.3 ozs. per 1,000 ft. of row for any row spacing (minimum 32-inch spacing)		
SORGHUM At Planting	Greenbugs	6.0 ozs. per 1,000 ft. of row for any row spacing (minimum 30-inch spacing)	Place granules in a 7-inch band over the row OR drill 1 inch below and 2 inches to the side of the seed.	Do not place THIMET granules in direct contact with seed. A second application of granules may be applied topically later in the season.
SORGHUM At Cultivation	Chinch bug nymphs (Colorado, Kansas, Nebraska)		Apply granules to base of plants or over the top of plants just ahead of cultivation shovels so as to cover granules with soil. Apply as soon as chinch bug nymphs are observed at base of plants. Granules must be placed at the infested region of the plants and adequately covered with soil.	Later infestations of chinch bugs may not be controlled. Do not make any application of THIMET after cultivation treatment. Do not feed foliage before grain harvest.
SORGHUM Over the Plant	Greenbugs Banks grass mites	4.9 lbs. per acre	Broadcast granules into the whorls of the plants by air or ground equipment.	Make only one application after plant emergence. Do not apply within 28 days of harvest.
SOYBEANS At Planting	For the early season control of: Mexican bean beetles Leafhoppers Lygus bugs Seedcorn maggots Thrips Aphids Mites	9.0 ozs. per 1,000 ft. of row for any row spacing (minimum 30-inch row spacing)	Drill granules to the side of the seed	Do not place THIMET granules in direct contact with seed. Do not feed the foliage of treated soybeans. Crop injury may result when preplant incorporated or preemergence applications of metribuzine herbicides are used in conjunction with this product or other soil-applied organophosphate pesticides.
	Seedcorn maggots		Place granules in a 7-inch band over the row directly behind the planter shoe and in front of the press wheel.	
SUGAR BEETS	Aphids Beet root maggots Leafhoppers Mites Leaf miners	4.5 ozs. per 1,000 ft. of row for any row spacing (minimum 20-inch spacing)	Drill granules to the side of the seed OR place granules in a band over the row.	Do not place THIMET granules in direct contact with the seed. Do not apply within 30 days of harvest. Do not feed sugar beet tops or silage to dairy cattle. Consult your state experiment station or state extension service for proper timing of application in your area.
SUGAR BEETS Postemergence	Mites Aphids	4.9-7.5 lbs. per acre	Apply to foliage when plants are dry. Use the higher rate on dense foliage for good coverage. Foliar application by air may be used for aphid control when wet soil prohibits ground application.	
SUGAR CANE (Florida)	Wireworms	19.5 lbs. per acre	Apply in a 10-12 inch band directly on the seed piece and surrounding soil in the open furrow immediately before covering.	
WHEAT At Planting	Grasshoppers (1) Aphids (Western and North Central States only) Hessian fly	1.2 ozs. per 1,000 ft. or row for any row spacing (minimum 8-inch spacing)	Apply the granules in the seed furrow.	Do not feed or graze foliage within 45 days of treatment. Do not make any later applications after planting time treatment. (1) Grasshoppers controlled in winter wheat only.
WHEAT Over the Plant	Aphids (Western States only)	4.9 lbs. per acre	Broadcast the granules evenly over the field by air equipment when aphids first appear.	Do not apply over plants if application was made at planting time. Do not make more than one application over the plants. Do not apply within 70 days of harvest of grain. Do not feed or graze foliage within 28 days of treatment.

4

Figure 12.1 Sample insecticide label, continued. (Courtesy American Cyanamid Company)

TYPICAL APPLICATION RATES
FOR VARIOUS ROW SPACINGS

Rate Oz./1,000 Ft. of Row

		1.2 oz.	3 oz.	4.5 oz.	6 oz.	8 oz.	10 oz.	12 oz.	14 oz.	16 oz.	18 oz.
	40"			3.7	4.9	6.5	8.2	9.8	11.4	13.1	14.7
	38"			3.9	5.2	6.9	8.6	10.3	12.0	13.7	15.5
	36"			4.1	5.4	7.3	9.1	10.9	12.7	14.5	16.3
	34"			4.3	5.8	7.7	9.6	11.5	13.5	15.4	17.3
EQUIVALENT POUNDS PER ACRE	32"			4.6	6.1	8.2	10.2	12.2	14.3	16.3	18.4
	30"			4.9	6.5	8.7	10.9	13.1	15.2	17.4	19.6
	26"			5.7	7.5	10.0	12.6	15.1	17.6	20.1	22.6
	24"			6.1	8.2	10.9	13.6	16.3	19.1	21.8	24.5
	20"		4.9								
	18"		5.4								
	16"		6.1								
	12"		8.2								
	8"	4.9	12.2								
	6"	6.5	16.3								

EQUIVALENT POUNDS PER ACRE

CALIBRATION INFORMATION

FIRST READ THE LABEL
It is important that applicator equipment be properly set to deliver the labeled rate. The table below contains suggested starting gauge setting for calibration at planting speeds of 5 or 7 mph and application rates of 4.5, 6.0 or 9.0 oz./1000 ft. of row.

NOTE: The settings in this table SHOULD ONLY BE USED AS STARTING POINTS. Continually check the amount of THIMET 20-G used against a known length of row and make further adjustments accordingly. Also check calibration occasionally to make sure equipment wear, changing moisture conditions, etc. have not caused a change in flow rate.

SUGGESTED GRANULAR APPLICATOR SETTING[1]						
	Planter Speed At 5 MPH			Planter Speed At 7 MPH		
Applicator	4.5 oz.	6.0 oz.	9.0 oz.	4.5 oz.	6.0 oz.	9.0 oz.
John Deere MaxEmerge 2	14	17	23	17	21	33
John Deere Max-Emerge (Odd Notches 5-15-25-35)	9	11	14	11	13	17
John Deere Max-Emerge (Even Notches 0-10-20-30)	9	13	20	13	19	26
John Deere 71 Flexi-Planter (John Deere Metal Hoppers)	1/23	1/26	2/3	1/26	1/32	2/14
New International Harvester (2 gauges/hopper)	1/9	2/2	2/6	2/2	2/5	3/3
Old International Harvester (1 gauge/hopper)	1/6	1/8	2/3	1/9	2/2	2/9
New Gandy	15	18	22	18	21	25
New Noble	6	8	11	8	11	14
Old Noble	8	11	16	12	15	22
Allis Chalmers (78 and 79 series)	4	5	7	4	5	7
Allis Chalmers #385	10	11	14	11	13	18

[1]Suggested starting point only.

DISCLAIMER

The label instructions for the use of this product reflect the opinion of experts based on field use and tests. The directions are believed to be reliable and should be followed carefully. However, it is impossible to eliminate all risks inherently associated with use of this product. Crop injury, ineffectiveness or other unintended consequences may result because of such factors as weather conditions, presence of other materials, or the use or application of the product contrary to label instructions, all of which are beyond the control of American Cyanamid Company. All such risks shall be assumed by the user.

American Cyanamid Company warrants only that the material contained herein conforms to the chemical description on the label and is reasonably fit for the use therein described when used in accordance with the directions for use, subject to the risks referred to above.

Any damages arising from a breach of this warranty shall be limited to direct damages and shall not include consequential commercial damages such as loss of profits or values or any other special or indirect damages.

American Cyanamid Company makes no other express or implied warranty, including any other express implied warranty of FITNESS or of MERCHANTABILITY.

In case of an emergency endangering life or property involving this product, call collect, day or night, Area Code 201-835-3100.

5

Figure 12.1   Sample insecticide label, continued. (Courtesy American Cyanamid Company)

⑮
ENDANGERED SPECIES RESTRICTIONS

The following restrictions apply to use of this product after February 1, 1988.

Before using this pesticide on corn, wheat, soybeans, sorghum, cotton, in the counties listed below, you must obtain the PESTICIDE USE BULLETIN FOR PROTECTION OF ENDANGERED SPECIES for the county in which the product is to be used. The bulletin is available from your County Extension Agent, State Fish and Game Office, or your pesticide dealer. **Use of this product in a manner inconsistent with the PESTICIDE USE BULLETIN FOR PROTECTION OF ENDANGERED SPECIES is a violation of Federal laws.**

ALABAMA—COLBERT, GREENE, JACKSON, LAMAR, LAUDERDALE, LIMESTONE, MADISON, MARSHALL, MORGAN, PICKENS AND SUMTER

ARIZONA—GRAHAM, MARICOPA, MOHAVE, PIMA, PINAL AND SANTA CRUZ

ARKANSAS—BENTON, CLAY, CLARK, CROSS, LAWRENCE, LEE, POINSETT, POLK, RANDOLPH, SHARP AND ST. FRANCIS

CALIFORNIA—BUTTE, COLUSA, GLENN, IMPERIAL, INYO, KERN, LOS ANGELES, MERCED, MODOC, ORANGE, RIVERSIDE, SACRAMENTO, SAN BERNARDINO, SAN DIEGO, SANTA BARBARA, SOLANO, STANISLAUS, SUTTER, TEHAMA, VENTURA AND YOLO

FLORIDA—ALACHUA, BAKER, BRADFORD, BREVARD, BROWARD, CHARLOTTE, CITRUS, CLAY, COLLIER, COLUMBIA, DADE, DE SOTO, DIXIE, DUVAL, FLAGLER, GADSDEN, GILCHREST, GLADES, HARDEE, HENDRY, HERNANDO, HIGHLANDS, HILLSBOROUGH, INDIAN RIVER, JEFFERSON, LAFAYETTE, LAKE, LEE, LEON, LEVY, MADISON, MANATEE, MARION, MARTIN, MONROE, NASSAU, ORANGE, OKEECHOBEE, OSCEOLA, PALM BEACH, PASCO, PINELLAS, POLK, PUTNAM, ST. JOHNS, ST. LUCIE, SARASOTA, SEMINOLE, SUMTER, SUWANNEE, TAYLOR, UNION, VOLUSIA AND WAKULLA

GEORGIA—BRANTLEY, BRYAN, BULLOCH, BURKE, CAMDEN, CANDLER, CHARLTON, CHATHAM, EFFINGHAM, EMANUEL, EVANS, GLASCOCK, GLYNN, JEFFERSON, JENKINS, JOHNSON, LIBERTY, LONG, McINTOSH, PIERCE, RICHMOND, SCREVEN, WARE, WASHINGTON AND WAYNE

KANSAS—CLARK, COMANCHE, MEADE AND STAFFORD

KENTUCKY—BALLARD, BUTLER, EDMONSON, GREEN, HART, JACKSON, LAUREL, LIVINGSTON, MARSHALL, McCRACKEN, McCREARY, PULASKI, ROCKCASTLE, TAYLOR, WARREN AND WAYNE

MISSISSIPPI—CLAIBORNE, COPIAH, HINDS, ITAWAMBA, LOWNDES, MONROE AND NOXUBEE

MISSOURI—BARRY, BENTON, CAMDEN, CHRISTIAN, DALLAS, GREENE, HICKORY, JASPER, LAWRENCE, MILLER, NEWTON, OSAGE, POLK, ST. CLAIR, STONE AND WEBSTER

MONTANA—GARFIELD, McCONE, SHERIDAN AND VALLEY

NEBRASKA—BOYD, BROWN, BUFFALO, BUTLER, CASS, CEDAR, COLFAX, DAWSON, DODGE, DOUGLAS, HALL, HAMILTON, HOLT, HOWARD, KEARNEY, KEYA PAHA, KNOX, MERRICK, NANCE, PHELPS, PLATTE, POLK, ROCK, SARPY AND SAUNDERS

NEVADA—CLARK

NEW MEXICO—CHAVES, DE BACA AND EDDY

NORTH CAROLINA—EDGECOMBE, NASH AND PITT

NORTH DAKOTA—BENSON, BOTTINEAU, BURKE, BURLEIGH, DIVIDE, DUNN, EDDY, EMMONS, FOSTER, KIDDER, LOGAN, McHENRY, McINTOSH, McKENZIE, McLEAN, MERCER, MORTON, MOUNTRAIL, NELSON, OLIVER, PIERCE, RAMSEY, RENVILLE, ROLETTE, SHERIDAN, SIOUX, STUTSMAN, TOWNER, WARD, WELLS AND WILLIAMS

OHIO—PICKAWAY

OKLAHOMA—DELAWARE, McCURTAIN AND PUSHMATAHA

OREGON—LAKE

SOUTH CAROLINA—AIKEN, BARNWELL, BEAUFORT, BERKELEY, CHARLESTON, COLLETON, DORCHESTER, GEORGETOWN, HAMPTON, HORRY, JASPER AND MARION

SOUTH DAKOTA—CLAY, HAAKON, HUGHES, POTTER, STANLEY, SULLY, UNION, WALWORTH, YANKTON AND ZIEBACH

TENNESSEE—BEDFORD, BLOUNT, CLAIBORNE, DECATUR, FRANKLIN, HANCOCK, HARDIN, HAWKINS, HICKMAN, KNOX, LAWRENCE, LINCOLN, LOUDON, MARSHALL, MAURY, MEIGS, MONROE, RHEA, ROANE, SCOTT, SEQUATCHIE, SMITH, SULLIVAN, TROUSDALE AND WAYNE

TEXAS—ARANSAS, AUSTIN, BASTROP, BURLESON, CAMERON, COLORADO, COMAL, FORT BEND, GOLIAD, HARRIS, HAYS, JEFF DAVIS, PECOS, REEVES, REFUGIO AND VICTORIA

UTAH—UTAH AND WASHINGTON

VIRGINIA—LEE, RUSSELL, SCOTT, SMYTH, TAZEWELL, WASHINGTON AND WISE

CYANAMID

American Cyanamid Company
Agricultural Division
⑯ Crop Protection Chemicals Department
Wayne NJ 07470   © 1987

6

Figure 12.1 Sample insecticide label, continued. (Courtesy American Cyanamid Company)

Labels and Labeling

The insecticide **label** is the printed information attached to and accompanying an insecticide container. This label is many things to many people. To the manufacturer, it is a "license to sell." To the state or federal government, it is a way to control the distribution, storage, sale, use, and disposal of the products. To the buyer or user, it is the source of facts relating to proper and legal product application. It is also a means to tell users about any special precautions needed in handling the product. Additionally, the label is a legal document; it is illegal to use an insecticide in any way that violates the instructions on the label.

The information contained on the label is based on scientific research conducted to provide facts necessary for product registration. Information on physical and chemical properties, toxicological profile, residue analysis and exposure estimates from projected uses, as well as product efficacy, must be submitted to EPA in order to receive product registration.

The insecticide label must conform to a set of standards established by federal law. Some labels are easy to understand, others are complicated. But, all labels will tell you how to use the product correctly, and all must contain the following information (illustrated on the specimen label, Fig. 12.1):

(1) **Trade name**. This is the brand name, used in advertising for the product. It usually shows up plainly on the front panel of the label (e.g., Thimet).

(2) **Type of formulation**. Somewhere on the label the type of formulation in the package will be described. Sometimes, the name clearly identifies the type of formulation (for instance, Thimet 20G indicates a formulation of 20% a.i. on granules), in other cases, the reader must search the label carefully in order to determine formulation type.

(3) **Ingredient statement**. This lists the constituents of the product. The amount of each active ingredient is given as percentage by weight (e.g., 20% active ingredient, 80% inert ingredients) or as pounds per gallon of concentrate.

(4) **Common name and/or chemical name**. Many insecticides have complex chemical names and some have been given common, or generic names as well to make them easier to identify. For example, phorate is the common name for 0,0- diethyl S-[(ethylthio)methyl]phosphorodithioate.

(5) **EPA registration and establishment numbers**.

(6) **Signal words and symbol**. Insecticides are toxic substances. The **signal word** and symbol give a rough estimate of the toxicity category of a product (see Table 12.1) or environmental hazard. Signal words include: danger-poison (accompanied by skull and crossbones symbol), warning, and caution. All labels must bear the statement: "keep out of reach of children".

(7) **Statement of practical treatment**. This gives instructions in case of accidental exposure. Additionally, it includes information for physicians and may provide notes on antidotes, if they are available.

(8) **Net contents**. This tells how much product (for instance, pounds, kilograms, gallons, liters) is in the container.

(9) **Statement of use classification**. Insecticide labels must show whether contents are for general-use or restricted-use. EPA classifies an insecticide as restricted-use if it could cause some human injury or environmental damage, even when used as directed on the label. Applicators must be certified by their state in order to apply restricted-use insecticides.

(10) **Precautionary statements**. These outline possible hazards to humans and domestic animals, environmental hazards, and physical and chemical hazards.

(11) **Reentry statement**. If required for the product, this section tells how much time must pass before a treated area is safe for entry by a person without protective clothing.

(12) **Storage and disposal instructions**. This section indicates where the product can be safely stored and how to dispose of excess material.

(13) **Directions for use**. The instructions on how to use the insecticide are a very important part of the label. They are the best way to determine the proper and legal ways to apply the material. As is evident in our specimen label (Fig. 12.1), these instructions can be long and quite complex. Many terms are used on the label that tell the user when and how to apply the insecticide (e.g., postemergence, sidedress, in-furrow). Appendix 12.3, at the end of this chapter, lists and defines some of the more common application terms. Additionally, this section lists the **harvest intervals** for various crops, the minimum lengths of time after insecticide applications, before crops can be harvested.

(14) **Misuse statement**. This section reminds the user that it is a violation of federal law to use an insecticide in a manner inconsistent with its labeling.

(15) **Endangered species restrictions**. This section lists counties by state where endangered species are present and special restrictions to use of the insecticide apply.

(16) **Manufacturer's name and address**.

The label is obviously a valuable reference source for the insecticide user before, during, and after insecticide application. Before an insecticide is purchased, the label should be read to determine: (1) whether the insecticide is needed for the job, and (2) whether the insecticide is too hazardous to be used safely under the application conditions to be employed. Before the insecticide is mixed, the label must be read to determine: (1) what protective equipment should be used, (2) what the insecticide can be mixed with (compatibility), (3) how much insecticide to use, and (4) the mixing procedure to use. Before the insecticide is applied, the label must be read to determine: (1) what safety measures should be followed, (2) where the insecticide can be used (livestock, crops, structures, etc.), (3) when to apply the insecticide (including the waiting periods for crops and animals), (4) how to apply the insecticide, and (5) whether there are any restrictions on the use of the material. Before insecticides or insecticide containers are stored or disposed of, the label must be read to determine: (1) where and how to store the material, (2) how to decontaminate and dispose of the empty insecticide container, and (3) where to dispose of any surplus insecticide.

Using Insecticides Safely

Most pesticides, including the insecticides, can cause severe illness, or even death, if they are misused. A product's **hazard**, or the danger that injury will occur, depends on the toxicity of the active ingredient plus the exposure to the product during use.

Children under ten years old are the victims of at least half of the accidental pesticide deaths in the US. If pesticides were always cared for properly, children would never touch them, and these deaths would be prevented. Most accidental pesticide deaths are caused by eating or drinking the product (often after leftover product has been improperly placed into an empty food container). Applicators are sometimes injured or killed when they breath a pesticide vapor or get a pesticide splashed on their skin.

Table 12.2 Toxicity categories of pesticides.

Measure of Toxicity	Highly Toxic	Moderately Toxic	Slightly Toxic	Rel. Nontoxic
Oral LD_{50}[1] (mg/kg)	0-50	50-500	500-5000	>5000
Dermal LD_{50} (mg/kg)	0-200	200-2000	2000-20000	>20000
Inhalation LC_{50}[2] (ppb)	0-200	200-2000	2000-20000	>20000
Lethal Dose, 150 lb. person	few drops to 1 tsp	1 tsp to 1 oz	1 oz to 1 pt	1 pt
Eye Effects	corrosive	irritation for 7 days	irritation for <7 days	none
Skin Effects	corrosive	severe irritation	moderate irritation	mild irritation
Signal Word/Symbol	DANGER-POISON w/ skull and crossbones	WARNING	CAUTION	CAUTION

[1] LD_{50} = The dose of a toxicant that will kill 50 percent of the test organisms to which it is administered; generally expressed as milligrams toxicant per kilogram of body weight.

[2] LC_{50} = The concentration of a toxicant in some medium (e.g., air) that will kill 50 percent of the test organisms exposed; generally expressed as mg or cm^3 per animal or as parts per million (ppm) or billion (ppb) in the medium.

Absorption through the skin, or **dermal exposure**, is the most common route of poisoning of agricultural workers. Absorption may occur as the result of a splash, spill, or drift when mixing, loading, applying, or disposing of pesticides. The degree of dermal absorption hazard depends on the dermal toxicity of the pesticide, the extent of the exposure, the type of formulation, and the source of contamination. Usually, wettable powders, dusts, and granular insecticides are not as readily absorbed through the skin and other body tissues as are liquid formulations such as the emulsifiable concentrates. Rates of absorption through the skin are different for different parts of the body (see Figure 12.2). Absorption through the skin in some regions (e.g., scrotal area) can be so rapid as to approximate the effect of injecting the material directly into the bloodstream. Absorption through the skin can take place as long as it is in contact with the insecticide. The seriousness of the exposure is increased if the contaminated area is large, or if the material remains on the skin for a long period of time.

Insecticide ingestion, or **oral exposure** may result in serious illness, severe injury, or even death. Insecticides may be consumed by accident, through carelessness, or they may be consumed intentionally. Oral exposure can be prevented by: (1) storing insecticides in their original, labeled containers, (2) avoiding all siphoning or clearing of lines or nozzles by the mouth, and (3) washing thoroughly after working with insecticides and before eating, drinking, or smoking.

The inhalation of insecticides, or **respiratory exposure**, is particularly hazardous because insecticides can be rapidly absorbed by the lungs. Additionally, insecticides may be inhaled in sufficient amounts to cause serious damage to nose, throat, and lung tissues. Vapors and extremely fine particles pose the most serious risks. The risks of respiratory poisoning usually are low when diluted sprays are applied with conventional application equipment, primarily because of the relatively large droplet size. When low-volume equipment is being used to apply concentrated material, the risk is increased substantially because of the smaller droplets produced. And there is significant risk when mixing and loading dust or powder formulations. Application in confined spaces (e.g., during treatment of stored grain) is particularly hazardous. Protection from respiratory exposure is provided by gas masks and other respiratory devices.

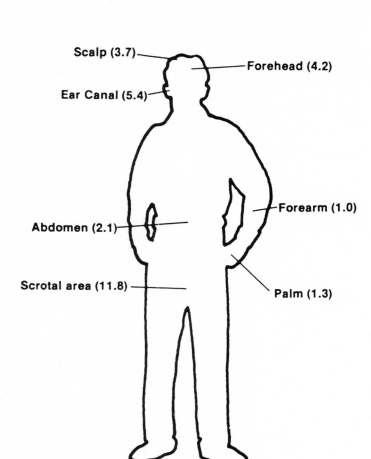

Scalp (3.7)

Forehead (4.2)

Ear Canal (5.4)

Abdomen (2.1)

Forearm (1.0)

Scrotal area (11.8)

Palm (1.3)

Ball of foot (1.6)

Figure 12.2 Relative dermal absorption rates of pesticides. (Courtesy Iowa Cooperative Extension Service)

The tissues of the eye are particularly absorbent and therefore susceptible to insecticide injury. Eye protection is needed when measuring or mixing insecticide concentrates and when applying highly or moderately toxic materials. Protective shields or goggles also should be used whenever there is a chance that sprays or dusts may come in contact with the eyes.

Recognizing the potential hazards of some insecticides, the responsible applicator will take every precaution to avoid adverse health effects. Keeping exposure levels to a minimum is the key to reducing risk. When using any pesticide, regardless of its toxicity, the applicator should wear at least a wide-brimmed hat, long-sleeved shirt, trousers or a coverall garment, underwear, socks, and shoes. Many times it is advisable to wear coveralls over regular clothing. When handling pesticide concentrates during mixing and loading or when using highly or moderately toxic materials, applicators should also wear unlined rubber gloves and boots, a rubber or vinyl apron, and goggles or a face shield. Pant legs should be worn outside of boots and sleeves over gloves to prevent pesticides from getting inside in case of a spill.

Protective gear and clothing should be changed and washed daily. If clothing gets wet with spray, change it promptly. If it gets contaminated with insecticide concentrates or highly toxic insecticides, destroy it, as it is difficult to clean by normal laundering methods.

Since insecticides can be absorbed through the skin, it is important to shower thoroughly after working with them. Hands should be washed after even incidental contact with insecticides and especially prior to eating, drinking, smoking, chewing tobacco, or going to the restroom. Applicators must avoid rubbing their eyes or touching their faces during or after insecticide application. Of course, no one should eat or store food in areas where insecticides or other pesticides are applied or stored.

INSECTICIDE APPLICATION

Choosing the appropriate application equipment is integral to the success of an insecticide application. All equipment must be properly selected, employed, and maintained for successful and safe insecticide applications. Many types of equipment are available. Dry insecticide materials may be applied with hand dusters, power dusters, and granular applicators. Liquid insecticide formulations are applied with hand sprayers, low pressure field sprayers, high pressure sprayers, air blast sprayers, ultra low volume (ULV) sprayers, aerosol generators, and foggers (a summary of insecticide application tools and their uses, advantages, and limitations is listed in Appendix 12.4 at the end of this chapter).

Proper selection of application equipment parts and accessories also should be practiced; sprayer tanks, pumps, filters, hoses, agitators, and nozzles are available in a variety of materials and capacities, and many are designed for specific uses. Choosing the proper sprayer nozzle type and size is very important. The nozzle determines the amount of spray applied to a particular area (see the following sections on calibrating field sprayers for more information), the uniformity of the applied spray, the coverage obtained on the sprayed surfaces, and the amount of drift. Nozzles and nozzle parts are available in several materials including brass, plastic, stainless steel, aluminum, and tungsten carbide. These materials all vary in their costs and susceptibilities to abrasion and corrosion.

Calibration

Calibration is the process of measuring and adjusting the amount of insecticide that application equipment will deliver to the target area. Proper calibration is an essential but often neglected task. It is important to apply the correct amount of insecticide: too little can result in inadequate insect pest control, and too much can cause injury to the target plant, animal, or surface, and may result in illegal residues, excess runoff or other movement from the target, injury to nontarget organisms, and resulting lawsuits and fines.

Before equipment is calibrated, it should be checked carefully to assure that all components are clean and in good working order. Various types of application equipment differ in the details of their operation, but if you understand the basic principles of calibration, you can apply them in any situation. Study the application equipment manufacturer's instructions carefully: they explain exactly how to adjust the equipment. Often the instructions contain suggestions on the appropriate driving speed, the range of most efficient pump pressures, approximate settings for achieving various delivery rates, and types of nozzles that can be used.

Calculating Insecticide Concentrations

To apply insecticide safely and effectively, an understanding of the recommended rate is essential. Usually, the insecticide label will tell you how much formulated material to apply per unit area. However, recommendations are sometimes stated as amount of active ingredient to apply per unit area. In this latter instance, the first step in the calibration process is to calculate how much formulated, commercial product (as poured from the bag or can) is needed to achieve the recommended or desired application rate.

Concentrations of dry insecticides, such as granules and wettable powders, are usually expressed as percentage of active ingredient (on a weight basis), whereas concentrations of liquids are generally expressed as number of pounds of active ingredient per gallon of formulated material. The following example calculations illustrate the common ways of determining the amounts of formulated materials needed to achieve a specific application rate.

Example 1
It is recommended that a 20% granular insecticide formulation be applied at a rate of 0.5 pounds of active ingredient per acre. How much formulated material must be applied per acre to achieve this rate?

	(0.5 lb active ingredient desired per acre)
divided by:	(0.20 proportion active ingredient in formulated product)
equals:	2.5 lb formulated product must be applied per acre

Example 2
You wish to mix 100 gallons of a spray mixture that will be 0.06% active ingredient. The formulated product is a 50% wettable powder. How many pounds of formulated product must you add to your spray tank? Note that water weighs 8.3 lbs per gallon.

	(100 gal of mixture x 0.0006 proportion active ingredient desired x 8.3 lbs/gal)
divided by:	0.50 proportion active ingredient in formulated product
equals:	0.996 or ca. 1 lb formulated material must be applied per acre

Example 3
How many gallons of a liquid insecticide formulation containing 4 pounds active ingredient per gallon are needed to apply 1.5 pounds active ingredient per acre?

	(1.5 lbs active ingredient/acre desired)
divided by:	4 lbs/gal active ingredient in formulated product
equals:	0.375 gal of formulated product must be applied per acre

Example 4
How many gallons of an emulsifiable concentrate insecticide formulation containing 3 pounds active ingredient per gallon are needed to make 150 gallons of a spray mixture containing 0.5% active ingredient?

	(150 gal of mixture x 0.005 active ingredient desired x 8.3 lbs/gal)
divided by:	3.0 lbs/gal active ingredient in formulated product
equals:	2.075 gal of formulated product must be added to 150 gal of water

Example 5
If 1 pound active ingredient of a 20% granular insecticide formulation per 13,068 linear feet of

row is recommended, how many ounces of formulated product should be delivered from each hopper of a granular applicator in 817 feet of row? Note that there are 16 ounces in one pound.

> 1 lb active ingredient desired per 13,068 feet
> divided by: 0.20 proportion active ingredient in formulated product
> to give: 5 lb formulated product in 13,068 feet of row
>   5 lb fomulated product/13,068 feet of row = .000383 lb/ft
>   .000383 lb/ft x 817 ft = .313 lb formulated product in 817 feet of row

You will note that in the above examples, English rather than metric units have been used in the calculations. Although widespread employment of the metric system is a desired goal in the US, application rates suggested on insecticide labels and in most Cooperative Extension Service recommendations are still given in the English system, and the English units (gallons, ounces, feet, acres, miles per hour, etc.) are well accepted and understood by the majority of growers and applicators. Of course, the same calibration logic applies regardless of the units employed. Conversion equations for the two systems are detailed in Appendix 12.5 at the end of this chapter.

Calibrating Field Sprayers

Once the desired amount of formulated material has been determined, the application equipment is ready to be calibrated. To calibrate a sprayer for liquid insecticide mixtures, one must bear in mind that three variables affect the amount of spray mixture applied to a unit area: (1) the nozzle flow rate, (2) the ground speed of the sprayer, and (3) the effective sprayed width per nozzle. For proper calibration and operation of a sprayer, the influence of these variables on sprayer output must be known and understood.

The flow rate through a nozzle varies with the size of the nozzle opening and the nozzle pressure. Installing a nozzle tip with a larger orifice or increasing the pressure will increase the flow rate. Nozzle flow rate varies in proportion to the square root of the pressure. Doubling the pressure does not double the flow rate. To double the flow rate, the pressure must be increased by four times (see Fig. 12.3). For example, to double the flow rate of a nozzle from 0.28 gallons per minute (GPM) at 20 pounds per square inch (PSI) to 0.56 GPM, the pressure must be increased to 80 PSI (4 x 20 PSI). Pressure cannot be used to make major changes in application rate but can be used to correct for minor changes due to nozzle wear. To obtain a uniform spray pattern and to minimize drift, the operating pressure must be kept within the range recommended for each nozzle type.

Spray application rate varies inversely with ground speed. Doubling the ground speed of the sprayer reduces the gallons of spray applied per acre by one-half (see Fig. 12.4). For example, a sprayer applying 20 gallons per acre (GPA) at three miles per hour (MPH) would apply 10 GPA if the ground speed was increased to six MPH and the pressure remained constant.

The effective width sprayed per nozzle also affects the spray application rate. Doubling the effective sprayed width per nozzle decreases the gallons per acre by one-half (see Fig. 12.5). For example, if 40 GPA is being applied with nozzles 20 inches apart, and if nozzles are changed, with the same flow rate, to 40 inches apart, the application rate decreases from 40 GPA to 20 GPA. Nozzle tip selection depends on the desired application rate in gallons per acre, ground speed (MPH), and the effective spray width per nozzle. The best way to choose

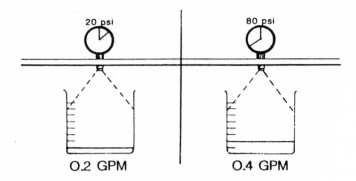

Figure 12.3 Relationship between flow rate and pressure. (Courtesy Iowa Cooperative Extension Service)

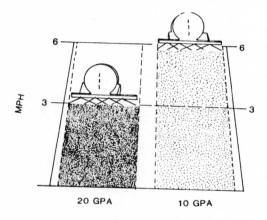

Figure 12.4 Relationship between spray application rate and ground speed of applicator. (Courtesy Iowa Cooperative Extension Service)

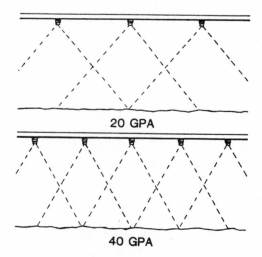

Figure 12.5 Relationship between spray application rate and effective spray width per nozzle. (Courtesy Iowa Cooperative Extension Service)

the correct nozzle tip is to determine the GPM required for your conditions and then select nozzles that provide this flow rate when operated within the recommended pressure range. The flow rate required from each nozzle in GPM can be determined by the following equation:

$$(GPA \times MPH \times W)/5940 = GPM$$

where:

GPM = gallons per minute of output required from each nozzle;

GPA = spray application rate in gallons per acre (from insecticide label);

MPH = ground speed of application equipment;

W = the effective spray width per nozzle (for broadcast spray: the nozzle spacing; for band spraying: the band width); and

5940 = a constant for converting gallons per minute, miles per hour, and inches to gallons per acre

Now a nozzle can be selected that gives the flow rate (GPM) determined above when the nozzle is operated within the recommended pressure range. Nozzle catalogs can be obtained from equipment dealers or nozzle manufacturers; the catalogs will list the GPM's at various pressures for different nozzles and this information will assist you in your selection.

Example 6

You want to broadcast an insecticide at 10 GPA at a speed of 7 MPH, using nozzles spaced 40 inches apart on the boom. What output in gallons per minute (GPM) would you expect under these conditions?

10 gal per acre desired x 7 mph ground speed x 40 inch nozzle spacing

divided by: 5940

equals: 0.47 gallons per minute

You would want to select a nozzle with a flow rate capacity of 0.47 GPM when operated within the recommended pressure.

Once the proper nozzle tips are selected and installed, sprayer calibration can be completed. The calibration should be checked every few days during the season and when changing the pesticide being applied. New nozzles do not lessen the need to calibrate because some nozzles "wear in", and will increase their flow rate rapidly during the first few hours of use.

There are many ways to calibrate equipment. The preferred methods differ according to the kind of equipment used. The following method, however, permits calibration in a minimum amount of time for either broadcast or band applications.

Broadcast or Band Sprayer Calibration

Step 1

Fill the sprayer tank at least half-full with water. This will simulate actual spraying conditions.

Step 2

Determine the nozzle spacing (for broadcast spraying) or the band width in inches and measure off the appropriate distance in the field according to Table 12.2.

Step 3

Select the tractor gear, and mark the throttle or speedometer setting to be used during the spraying operation. Start the tractor 25 feet behind the starting point of the distance marked off in Step 2. Start timing your travel time in seconds as you cross the starting point, stop timing when you cross the end point of the designated distance.

Table 12.2   Travel distances for calibrating sprays at different nozzle spacings or band widths.

Broadcast nozzle spacing or band width (inches)	Travel distance (feet)
5	538
14	291
20	204
30	136
36	113
38	107
40	102

Step 4
Set the pressure at the reading to be used to spray the field.
Step 5
With the sprayer stationary and with water in the sprayer tank, collect water from each nozzle at the pre-set pressure (set in Step 4) for the number of seconds it took to travel the prescribed distance (from Steps 2 & 3). The quantity of water delivered by each of the nozzles should be nearly the same. If a nozzle's delivery varies by more than 5% to 10% from the delivery of the other nozzles on the boom, it should be replaced.
Step 6
The number of ounces of water collected per nozzle equals the number of gallons per acre delivered by the sprayer: ounces per nozzle = gallons per acre.
Step 7
Next, determine the amount of insecticide to add to the spray tank. To make this calculation you need to know the recommended application rate, the capacity of the spray tank, and the output of the sprayer (calibrated in the above steps 1-6).

Example 7
You wish to apply a broadcast application of an insecticide at a rate of 3 quarts per acre. The sprayer has a 250 gallon tank and is calibrated to apply 20 gallons per acre. How many acres can be sprayed with each tankful?

	tank capacity of 250 gallons
divided by:	sprayer calibrated to apply 20 gal per acre
equals:	12.5 acres can be treated with each tankful

How much insecticide must be added to each tankful?

	12.5 acres per tankful x 3 quarts per acre
equals:	37.5 quarts of insecticide must be added to each tankful

Example 8
A farmer has corn rows that are 38 inches apart. He wants to apply an insecticide at the rate of 2 quarts per acre, but in a 14 inch band, to 120 acres. His sprayer is calibrated to apply 30 GPA and his tank's capacity is 300 gallons. How much area (in acres) actually need to be

treated with insecticide?

	120 acres of corn x 14 inches per band
divided by:	38 inches per row
equals:	44.2 acres need to be treated

How much insecticide does he need to treat this area?

44.2 acres x 2 quarts of insecticide per acre = 88.4 quarts of insecticide

How many acres can be sprayed with each tankful?

	tank capacity of 300 gallons
divided by:	sprayer calibrated to apply 30 gallons per acre
equals:	10 acres can be treated with each tankful

How much insecticide must be added to each tankful?

	10 acres per tankful x 2 quarts per acre
equals:	20 quarts of insecticide must be added to each tankful

Calibrating Granular Applicators

The calibration of equipment designed for applying granular pesticide formulations depends on the size of the metered openings of the applicator, the speed of the agitators or rotors, travel speed, the roughness of the field, and the flowability of the granules. Granules flow at different rates, depending on their size, density, and composition, as well as on temperature and humidity. A different applicator setting may be necessary for each pesticide applied as variations in flow rate may occur with the same product from day to day or from field to field. It is, therefore, very important to calibrate frequently in order to maintain the proper application rate.

Besides the setting of the metered opening, ground speed is the most significant factor influencing the application rate. The ground speed during calibration and application must be the same, and the speed must remain constant. A speed change of only one mile per hour can cause a significant variation in the application rate.

The method chosen for calibrating a granular applicator depends largely on the type of equipment and method of application, as well as the way in which the application rate is expressed. The following method is simple and will permit calibration for granular insecticides in a seven-inch band width or for in-furrow applications.

Granular Applicator Calibration

Step 1
Measure and mark a distance of 1000 feet in the field.
Step 2
Examine the insecticide label to determine how many ounces of insecticide are to be applied per 1000 feet of row.
Step 3
Fill the hoppers on the granular applicator with the same granules you intend to use during the application, turn the applicator on, and operate until all the hoppers are feeding. Turn the the applicator off, disconnect the drop tubes, and attach a container (e.g., a plastic bag or plastic

jar) to the outlet from each hopper. Be sure to record the weights of the containers used.

Step 4

Drive the measured distance with the applicator turned on and the containers collecting the output from each hopper. Be sure to operate at the speed that will be used during the actual application.

Step 5

Weigh and record the amount of material collected from each hopper (be sure to subtract the container weights from the total weights). Compare the weight of the collected insecticide to the recommended application rate. Adjust the setting on every hopper unit that is not within 5% of the recommended rate and then recalibrate. Note that it is not unusual for the settings to be different on different hoppers.

Example 9

If 8 ounces of formulated product are recommended per 1000 linear feet of row, how many ounces per hopper should be collected if rows are 36, 38, or 40 inches apart?

Answer:

8 ounces per hopper under any of the row widths: row width can be ignored as long as you know how many ounces are required per 1000 feet of row. Most of today's recommendations are based on 1000 feet of row due to the great variation in row spacings employed by different growers.

In summary, it is important to remember that the success or failure of insecticidal management of pests often depends upon the suitability of the equipment for the particular situation and the care and skill with which the operator uses the equipment. Two basic objectives in operating insecticide application equipment include: (1) accurate application of the desired rate, and (2) uniform distribution of the insecticide. The purpose of calibration is to insure that the application equipment is ready to do these two jobs.

OBJECTIVES

1) Know the primary routes of insecticide entry and modes of action.

2) Familiarize yourself with the common types of insecticide formulations and their properties, advantages, and disadvantages.

3) Be able to identify and understand the parts of an insecticide label.

4) Be familiar with the toxicity categories of insecticides and know how to protect yourself against poisoning.

5) Know how to calibrate a field sprayer and a granular applicator and be prepared to perform any necessary calculations related to calibration.

6) Additional objectives (at instructor's discretion):

LIST OF EXERCISES

1) Calibration Problems. Make calculations for sample calibration problems. Instructions on page 254 (answers on page 255).

OPTIONAL DISPLAYS AND EXERCISES

1) Examine the sample insecticide labels on display. Identify the trade name, formulation, common or chemical name, signal words, etc. Compare a label from a general-use insecticide to that of a restricted-use product.

2) Examine the pesticide application equipment on display. How can the delivery rates of each piece be altered? Attempt to calibrate a piece of equipment to deliver a specific amount of material per unit of application area. Be sure to handle the water, inert dust, and granules as if they are genuine insecticides.

3) Examine the protective equipment and clothing on display. What cleaning and maintenance procedures would be used for each piece? What are the advantages and disadvantages of each type of protective gear?

4) Using any resource materials available to you, seek out insecticide application recommenda-tions for one or more of the following insect pest problems. Evaluate the recommendations in terms of hazard to the user/applicator, hazard to the environment and to nontarget organisms, ease of mixing and application, and reentry time or persistence of residues. Speculate on possible non-insecticidal tactics for dealing with the insect pests.
     a. grasshoppers in grass pasture
     b. Indianmeal moth in a grain bin filled with whole kernel corn
     c. cattle lice on lactating dairy cattle
     d. German cockroaches in a school cafeteria
     e. mimosa webworms on a homeowner's honey locust trees
     f. sod webworms in golf course turf
     g. fleas on a pet cat
     h. asparagus beetles on a commercial grower's asparagus plants
     i. whiteflies in the greenhouse

5) Additional displays and exercises (at instructor's discretion):

TERMS

active ingredient
aerosol
bait
broad-spectrum
calibration
contact poison
dermal exposure
dust formulation
emulsifiable
    concentrate
flowable formulation
fumigant
granular formulation
harvest interval

hazard
hydrophilic
inert ingredient
insecticide
    formulation
label
LC_{50}
LD_{50}
lipophilic
metabolic inhibitor
microencapsulated
    formulation
mode of action
muscle poison

neurotoxin
nonselective
    insecticide
oral exposure
physical toxicant
replacement
rescue treatment
residual activity
resistance
respiratory exposure
resurgence
selective insecticide
signal word

soluble powder
    formulation
solution
sticker
stomach poison
surfactant
synapse
synergist
systemic
water-dispersible
    granules
wettable powder
    formulation

DISCUSSION AND STUDY QUESTIONS

1) What factors must be considered in selecting an insecticide to be used on vegetables intended for human consumption? On ornamental plantings in a public garden? On a forage crop that will be cut and fed to livestock at a later date?

2) Speculate on the advantages, disadvantages, and possible uses for insecticides possessing the following chemical or physical properties.
    a. high degree of water solubility
    b. long residual life, even in the presence of heat and water
    c. subject to photodegradation (i.e., light destroys the material)
    d. high volatility and strong odor
    e. high lipophilicity, tends to accumulate in fats
    f. is a steroid that can mimic the insect hormone ecdysone

3) In recent years the public has grown increasingly concerned with protecting groundwater supplies from contamination with pesticides and other manufactured chemicals. How might insecticides get into groundwater? What chemical and physical properties of an insecticide might favor entry into groundwater?

4) Write a press release for publication in a local newspaper or a public service announcement for broadcast on a local radio station that stresses the importance of insecticide application equipment calibration to growers in your area.

BIBLIOGRAPHY

Pedigo, L. P. 1989. Entomology and pest management. Macmillan Pub. Co., New York, NY
    - Chapter 11. Conventional insecticides.

Bode, L. E., and B. J. Butler. 1981. Equipment and calibration: low-pressure sprayers. Univ. of Illinois Coop. Extn. Serv. Cir. 1192
- a well-illustrated summary of sprayer types and their calibration.

Chemical and Pharmaceutical Press. 1986. Crop Protection Chemical Reference. 2nd ed. Chemical and Pharmaceutical Publishing Corp. and John Wiley and Sons, New York, NY
- a comprehensive reference which includes a complete listing of pesticide product information by manufacturer; appendices on calibration, pesticide handling and storage, poison control centers, solid and hazardous waste agencies; and indices of pesticide manufacturers and pesticide trade, common, and chemical names.

Entomological Society of America. Insecticide and Acaracide Tests. Entomological Society of America, College Park, MD
- published yearly, this journal is a compilation of reports of crop protection chemical efficacy. Reports are listed by type of commodity being protected.

Iowa State University Cooperative Extension Service. 1987. The Iowa Core Manual: A Guide for Commercial Pesticide Applicators. Publication IC-445
- this publication was the source for much of this chapter's information on insecticide safety, application, and calibration. Each US state has a comparable publication available, intended primarily for training of pesticide applicators. Other more specialized training manuals generally are available from each state's Extension Service as well (e.g., for applicators involved in aquatic pest control, regulatory pest control, termite control, forest pest control, etc.).

Matsumura, F. 1985. Toxicology of Insecticides. 2nd ed. Plenum Press, New York, NY
- a comprehensive survey of the toxicology of insecticides; includes discussion of insecticide chemistry, modes of action, and metabolism

Matthews, G. A. 1979. Pesticide Application Methods. Longman Group Ltd, New York, NY
- covers many technical aspects of pesticide application including detailed descriptions of ground and aerial pesticide application equipment.

Meister, R. T., ed. 1988. Farm Chemicals Handbook. 74th ed. Meister Publ., Willoughby, OH
- contains detailed descriptions of crop chemicals of all types. Because a new edition is released yearly, this publication is one of the most up-to-date resources available for persons interested in farm chemicals.

Ottoboni, M. A. 1984. The dose makes the poison. Vincente Books, Berkeley, CA
- an extremely well-written and informative book that explains the concepts of toxicity, poisoning, and carcinogenicity in a clear and thorough fashion. It is unfortunate that the information presented in this book is not more widely understood. This should be required reading for anyone using or making decisions about chemicals.

Society of Chemical Industry. Pesticide Science. Elsevier Sci. Publ., New York, NY

- a refereed journal of international research and technology on crop protection and pest control which includes technical articles covering: synthesis and screening of new pesticides; formulation, application, metabolism, degradation, toxicology, efficacy, and safety of new and existing pesticides, and ecological implications of pesticide application.

Ware, George W. 1983. Pesticides: Theory and Application. W. H. Freeman and Co., New York, NY

- a thorough examination of pesticide application, handling, storage, disposal, chemistry, mode of action, and biological activity. It contains a valuable set of appendices listing virtually all pesticides in current use, or restricted, suspended, and canceled.

Exercise I. Calibration Problems

1) A farmers's spray tank holds 250 gallons and is calibrated to apply 20 gallons per acre. How many acres can be treated with one tankful? How many gallons of emulsifiable concentrate insecticide containing 4 pounds of active ingredient per gallon of formulation should be added to the tank to apply 1.5 pounds active ingredient per acre?

2) A 50% wettable powder insecticide is to be applied at 1.5 pounds of active ingredient per acre. It will be applied in a 10-inch band over the row. How much formulated material is needed to treat 80 acres?

3) You want to broadcast an insecticide spray mixture at 20 gallons per acre at a speed of 5 mph using nozzles spaced 30 inches apart on the boom. What output in gallons per minute would you expect under these conditions?

4) If nine nozzles sprayed the following amounts in 3 minutes, which of the nine nozzles should be replaced?

Nozzle	Ounces/3 Minutes
1	28
2	32
3	29
4	35
5	31
6	30
7	28
8	26
9	31

5) If a field is being sprayed through nozzles 40 inches apart, and the average output per nozzle is 7 ounces per 102 feet of travel, how many gallons per acre are being applied? How many gallons would be applied to 80 acres?

6) Suppose you have a 300 gallon sprayer tank and your sprayer is calibrated to apply 25 gallons per acre. You are going to spray a 7-inch band over 30-inch corn rows. How many acres of the treated bands will you actually spray with one tankful? How many crop acres will you drive over in order to actually spray the one tankful?

7) Your sprayer is calibrated to apply 30 gallons per acre. The tank capacity is 500 gallons. How many fillings are required to spray a 160 acre field?

Calibration Problem Solutions

1)   (250 gal)/(20 gal/acre) = 12.5 acres per tankful
           12.5 acres/tankful x 1.5 lbs active ingredient/acre desired
divided by: 4 lbs active ingredient/gal
equal: 4.7 gal of formulated material per tank

2)   1.5 lbs active ingredient/acre
divided by: 0.50 proportion active ingredient in formulation
equals: 3 pounds formulated material needed per acre
           3 lbs formulated material x 80 acres to be treated
equals: 240 lbs formulated material is required

3)   (20 gallons/acre desired x 5mph x 30 inch spacing)/5940
equals: 0.5 gallons per minute

4)   The average nozzle output is 30 ounces and 5% of 30 is 1.5 ounces (10% of 30 is 3 ounces). You should definitely replace nozzles 4 and 8, which vary by 10% of the average and you should consider replacing 1, 2, and 7 as well because they vary by 5% from the average.

5)   7 gal per acre (because 7 ounces per 40 inches of boom equals 7 gal per field acre)

           7 gal/acre x 80 acres = 560 gal required

6)   (300 gal/tank)/(25 gal/acre) = 12 acres of treated bands per tankful
           (12 acres of bands/tankful x 30 inches/row)/(7 inches/band)
equals: 51.4 actual crop acres per tankful
(With each tankful, you will actually spray 12 acres of soil within the bands; however, since you are only spraying 7 inches of each 30 inches in the field, you will drive over 51.4 acres of crop in order to actually spray 12 acres of bands.)

7)   (500 gal/tank)/(30 gal/acre) = 16.7 acres per tankful
           (160 acres)/(16.7 acres/tankful) = 9.6 fillings of the tank

Appendix 12.1 - Federal Laws Regulating Pesticides in the US

1906 Federal Food, Drug, and Cosmetic Act (Pure Food Law)
- prevented the manufacture, sale, or transportation of adulterated, misbranded, poisonous, or deleterious foods, drugs, medicines, and liquors.

1910 Federal Insecticide Act
- prevented manufacture, sale, or transportation of adulterated or misbranded insecticides and fungicides

1938 Amendment: Pure Food Law
- amended to include pesticides on food; also required the adding of color to white insecticides to prevent their accidental use as cooking materials

1947 Federal Insecticide, Fungicide, and Rodenticide Act (FIFRA)
- superseded the 1910 Federal Insecticide Act and extended coverage to include herbicides and rodenticides
- required that pesticides be registered with the US Department of Agriculture and required good and useful labeling of pesticides (label was required to list name, brand, and trademark of the pesticide, its net contents and ingredients, the manufacturer's name and address, warning statements to prevent injury to humans, animals, plants, and nontarget invertebrates, and instructions for proper use)

1954 Miller Pesticide Residue Amendment to the Food, Drug, and Cosmetic Act
- provided that any raw agricultural commodity could be condemned as adulterated if it contained a residue of any pesticide whose safety had not been formally cleared or which was present in excessive amounts (above residue tolerances)

1958 Food Additives Amendment to the Food, Drug, and Cosmetic Act
- prescribed safe conditions for the use of food additives
- included the **Delaney clause** which stipulated that no material causing cancer under any condition could be permitted in food

1964 FIFRA amendment
- FIFRA was amended to require that: (1) all pesticide labels were to contain the federal registration number; (2) the front label of all pesticides was to include the words "warning", "danger", or "caution" and "keep out of reach of children"; and (3) all safety claims were to be removed from labels

1970 Environmental Protection Agency (EPA) established
- the responsibility for pesticide regulation was transferred from the USDA to EPA, and authority to establish pesticide tolerances on food was transferred from the Food and Drug Administration (FDA) to the EPA (however, enforcement of tolerances was to remain the responsibility of the FDA)

1972 Federal Environmental Pesticide Control Act (FEPCA) (also referred to as FIFRA Amended, 1972)

- provisions included:

    (1) use of any pesticide inconsistent with the label is prohibited;

    (2) deliberate violation of FEPCA can result in fines and/or imprisonment

    (3) pesticides must be classified as general-use or restricted-use

    (4) persons applying restricted-use pesticides must be certified by his or her state (includes commercial applicators and farmers)

    (5) pesticide manufacturers must be registered and inspected by the EPA

    (6) all pesticides must be registered by the EPA

    (7) states may register pesticides on a limited basis for Special Local Needs (SLN)

    (8) manufacturers must supply scientific evidence that a pesticide, when used as directed, will be effective, will not injure humans, crops, livestock, non-target organisms, or the environment, and will not produce illegal residues on or in food or feed

1972 Resource Conservation and Recovery Act

- gave EPA authority to control the disposal of hazardous wastes, including many pesticides

1973 Endangered Species Act (amendments: 1976, 1977, 1978, 1979, 1982)

- goal was to provide protection for threatened and endangered species: pesticide labels must be designed to protect specific endangered species from the adverse effects of pesticides and therefore, there are restrictions on pesticide application in endangered species' habitats

1975, 1978, 1980, 1981, 1988 FIFRA amendments

- provisions were included to clarify the intent of the law and to further influence pesticide registration and application; the 1988 amendment removed the requirement that the EPA indemnify (make compensation payments to) manufacturers for existing stocks of canceled chemicals

1986 Federal Emergency Planning and Community Right-to-Know Act (a provision of the Superfund Amendments and Reauthorization Act, SARA, of 1986)

- goal was to help increase the public's knowledge and access to information on the presence of hazardous chemicals, including pesticides, in their communities and releases of these chemicals into the environment

Appendix 12.2 - Common Insecticide Formulations

Aerosol (A)
Description: contains one or more active ingredients and a solvent; is driven through fine openings by gas under pressure, creating fine droplets
Advanatages: ready to use and convenient
Disadvantages: expensive; limited uses; inhalation risk exists; difficult to confine spray; hazardous if punctured, overheated, or used near an open flame

Bait
Description: consists of an edible or attractive substance mixed with insecticide; usually less than 5% active ingredient
Advantages: many can be used indoors; low hazard for applicator
Disadvantages: ideally, the attractiveness of the bait should be greater than that of the crop or other commodity to be protected

Dust (D)
Description: contains 1-10% active ingredient plus a very fine, dry inert carrier
Advantages: no mixing required (ready to use); effective in hard-to-reach indoor areas; requires simple equipment; valuable where moisture from spraying might cause damage
Disadvantages: high drift hazard; expensive

Emulsifiable Concentrate (EC or E)
Description: contains active ingredient, one or more petroleum solvents, and an emulsifier which allows the formulation to be mixed with water
Advantages: requires little agitation, leaves little visible residue;
Disadvantages: tends to be phytotoxic, easily absorbed through skin, and corrosive

Flowable (F)
Description: contains finely-ground active ingredients mixed with a liquid, along with inert ingredients, to form a suspension that is mixed with water
Advantages: seldom clogs nozzles; is easy to handle and apply
Disadvantages: may require moderate agitation and may leave a visible residue

Fumigant (F or LG)
Description: volatile liquid or solid that becomes a gas when applied
Advantages: can penetrate cracks, crevices, wood, and tightly-packed areas
Disadvantages: virtually no residual activity; target area must be enclosed or covered; highly toxic; application requires specialized protective equipment

Granular (G)
Description: active ingredient either coats or is absorbed by coarse, absorptive materials; the amount of active ingredient is usually 1-20%
Advantages: ready-to-use; low drift hazard; low applicator hazard; simple application equipment
Disadvantages: does not stick to foliage; somewhat expensive; may require incorporation into

the soil; may need moisture to activate insecticidal action

Microencapsulated
Description: consists of particles of insecticide surrounded by a permeable coating; is mixed with water and applied as a spray; once applied, the capsules slowly release insecticide
Advantages: easy and relatively safe to handle and apply
Disadvantages: constant agitation necessary; bees may pick up capsules and carry them to hives

Solution (S)
Description: dissolves readily in water and will not settle out or separate
Advantages: no agitation needed
Disadvantages: few solution formulations are available

Soluble Powder (SP)
Description: looks like a wettable powder but dissolves in water into a true solution; 15-95% active ingredient
Advantages: no agitation required
Disadvantages: inhalation hazard during mixing

Water-Dispersible Granules (or Dry Flowable)
Description: similar to wettable powders except that the active ingredient is prepared as granules that must be added to water for application
Advantages: easier and safer to measure and mix than wettable powders
Disadvantages: requires constant agitation and is abrasive to spray equipment

Wettable Powder (W or WP)
Description: dry, finely-ground powder that usually is mixed with water for application as a spray; contains 5-95% active ingredient and does not dissolve in water
Advantages: relatively inexpensive; easy to store; lower phytotoxicity than EC's, easily measured and mixed
Disadvantages: inhalation hazard while pouring and mixing concentrated powder; requires good and constant agitation and is abrasive to many pumps and nozzles; residues may be visible on foliage and surfaces

Appendix 12.4 - Insecticide Label Terminology

Many terms are used on the insecticide label to describe when and how to use insecticides. They also are found in leaflets and bulletins that you may get from your Cooperative Extension Service, land-grant university, or other agencies. Understanding these terms will help you get the best results from insecticide use.

Terms that tell you when to use the insecticide:

prebloom - used before the crop flowers
preplant - used before the crop is planted
preemergence - used before the crop or pests emerge; may also refer to use after crops emerge or are established, but before pests emerge
postemergence - used after the crop and pests have emerged

Terms that tell you how to use the pesticide product include:

band - application to a strip over or along a crop row or on or around a structure
basal - application to stems or trunks at or just above the ground line
broadcast - uniform application to an entire, specific area
crack and crevice - application in structures to cracks and crevices where pests may be harbored
dip - complete or partial immersion of a plant, animal, or object in an insecticide
directed - aiming the insecticide at a portion of the plant, animal, or structure
drench - saturating the soil with an insecticide; also the oral treatment of an animal with a liquid
foliar - application to the leaves of plants
in-furrow - application to the furrow in which a plant is planted
over-the-top - application over the top of the growing crop
pour-on - pouring the insecticide along the midline of the backs of livestock
sidedress - application along the side of a crop row
soil application - application to the soil rather than to vegetation

Appendix 12.4 - Insecticide Equipment

Hand Sprayer
Description: for application in structures, for small jobs, and for restricted areas where a power unit will not work
Advantages: economical; simple to use, clean, and store
Disadvantages: often lacks good agitation and screening needed for using wettable powders

Low-Pressure Field Sprayer
Description: delivers low to moderate volumes at low pressures for treating field and forage crops, pastures, fence rows, and buildings
Advantages: has medium to large tanks, low cost, light weight, and versatility
Disadvantages: low pressure limits insecticide penetration; low output limits use when high volume is required; agitation is limited

High-Pressure Field Sprayer (Hydralic Sprayer)
Description: delivers large volumes at high or low pressures; used to spray fruit trees, vegetables, trees, landscape plants, and livestock
Advantages: sturdy and long-lasting; has mechanical agitation
Disadvantages: high cost; needs large amounts of water, power, and fuel; high pressure spray is subject to drift

Air Blast Sprayer
Description: uses a high speed, fan-driven air stream to break nozzle output into fine droplets, which move with the air stream to the target; used to treat landscape plants, fruits, vegetables, and for biting fly control; applies either high or low volumes
Advantages: has good coverage and penetration, low pump pressures, and mechanical agitation
Disadvantages: may present drift problems and overdosages; it is hard to confine discharge to the target; is difficult to use in small areas

Ultra Low Volume Sprayer (ULV)
Description: contains special nozzles which spin at high speeds and break undiluted insecticides into uniformly-sized droplets by centrifugal force; droplets are carried to the target by gravity or by an air stream created by a fan
Advantages: does not need water, and control may be achieved with a small amount of insecticide
Disadvantages: does not provide for thorough wetting; concentrated insecticides are hazardous to handle; only a few insecticides may be used in this type of sprayer

Aerosol Generators and Foggers
Description: aerosols are created by using atomizing nozzles, spinning disks, and small nozzles at high pressure; fogs are generated by thermal generators using heated surfaces
Advantages: give efficient distribution of liquid insecticides in enclosed spaces and dense foliage
Disadvantages: aerosols and fogs are extremely sensitive to drift, and repeated applications are needed to maintain effectiveness

Hand Duster

Description: for use in structures and in gardens; consists of a squeeze bulb, bellows, shaker, sliding tube, or fan powered by a hand crank

Advantages: good penetration in confined spaces; fast and easy to use

Disadvantages: high cost for insecticide in this ready-to-use form; hard to get good foliar coverage; dust is subject to drifting

Power Duster

Description: uses a powered fan or blower to propel dust to a target

Advantages: simple construction, low cost, easy maintenance

Disadvantages: uniformity of application is not as good as with sprays; dust formulations are expensive; dusts are subject to drift

Granular Applicators

Description: may use spinning disks for broadcast coverage, or mounted equipment for applying bands over the row in row crops, or mounted or tractor-drawn machines for broadcast coverage

Advantages: no mixing needed; applicators are low in cost; present a low hazard to the applicator; low drift hazard

Disadvantages: granules are expensive, won't stick to plants, and receive poor lateral distribution (especially on slopes); must recalibrate for each granular formulation and whenever the humidity changes

Appendix 12.5 - Convenient Conversion Factors

MULTIPLY:	BY:	TO GET:
acres	43,560	square feet
acres	4840	square yards
acres	0.405	hectares
bushels	2150.42	cubic inches
bushels	32	quarts
centimeters	0.3937	inches
centimeters	0.01	meters
centimeters	10	millimeters
cubic feet	1728	cubic inches
cubic feet	0.03704	cubic yards
cubic feet	7.4805	gallons
cubic feet	59.84	pints
cubic feet	29.92	quarts
cubic inches	16.39	cubic centimeters
cubic meters	1,000,000	cubic centimeters
cubic meters	35.31	cubic feet
cubic meters	61,023	cubic inches
cubic meters	1.308	cubic yards
cubic meters	264.2	gallons
cubic yards	27	cubic feet
cubic yards	46,656	cubic inches
cubic yards	0.7646	cubic meters
cubic yards	202	gallons
degrees centigrade	1.8 (+ 17.98)	degrees Fahrenheit
degrees Farhenheit	0.5555 (- 32)	degrees centigrade
feet	30.48	centimeters
feet	12	inches
feet	0.3048	meters
feet	0.3333	yards
feet per minute	0.01667	feet per second
feet per minute	0.00508	meters per second
feet per minute	0.01136	miles per hour
gallons	3785	cubic centimeters (millimeters)
gallons	0.1337	cubic feet
gallons	231	cubic inches
gallons	128	ounces (liquid)
gallons	8	pints
gallons	4	quarts
gallons of water	8.3453	pounds of water
grams	0.001	kilograms
grams	1000	milligrams
grams	0.0353	ounces
grams per hectare	0.0143	ounces (dry) per acre
grams per liter	1000	parts per million

MULTIPLY:	BY:	TO GET:
inches	2.54	centimeters
inches	0.08333	feet
inches	0.02778	yards
kilograms	1000	grams
kilograms	2.205	pounds
kilograms per hectare	0.894	pounds per acre
kilograms per square centimeter	14.22	pounds per square inch
kilometers	3281	feet
kilometers	1000	meters
kilometers	0.6214	miles
kilometers	1094	yards
liters	1000	cubic centimeters (millimeters)
liters	0.0353	cubic feet
liters	61.02	cubic inches
liters	0.001	cubic meters
liters	0.2642	gallons
liters	2.113	pints
liters	1.057	quarts
meters	100	centimeters
meters	3.281	feet
meters	39.37	inches
meters	0.001	kilometers
meters	1000	millimeters
meters	1.094	yards
miles	5280	feet
miles	1760	yards
miles per hour	88	feet per minute
miles per hour	1.467	feet per second
miles per hour	1.609	kilometers per hour
milliliters per hectare	0.0137	ounces (liquid) per acre
ounces (dry)	28.3495	grams
ounces (dry)	0.0625	pounds
ounces (liquid)	1.805	cubic inches
ounces (liquid)	0.00781	gallons
ounces (liquid)	29.573	milliliters (cubic centimeters)
ounces (dry) per acre	0.07	grams per hectare
ounces (liquid) per acre	73.14	milliliters per hectare
pounds per acre	1.12	kilograms per hectare
pounds per square inch	0.0703	kilograms per square centimeter

13 BIOLOGICAL AND GENETIC CONTROL

Biological and genetic control are pest management approaches that employ other organisms and features of the pest itself to reduce pest numbers. The suitability of these tactics may be limited depending on the pest species, particularly with genetic control, but both tactics have the advantage of usually being environmentally sound. These tactics more often are employed through governmental agencies rather than by individual growers.

BIOLOGICAL CONTROL

Many insects are not major pests or are not pests at all because their populations are held in check through the action of **natural enemies**, organisms that kill insects or reduce their ability to survive and reproduce. Pest mortality through the action of natural enemies and other environmental factors is termed **natural control**. But the importance of natural enemies to pest management goes beyond their providing natural control; natural enemies are the basis for the management tactic of biological control. **Biological control** is the purposeful manipulation of natural enemies to produce a reduction in a pest's population. Some authorities include management tactics such as environmental manipulations, host plant resistance, or genetic control as part of biological control. However, we believe it is more appropriate to restrict biological control to considerations of natural enemies.

Unlike some other pest management tactics, biological control usually poses no threats to the environment and may be a self-perpetuating technique once established. Indeed, in certain instances biological control may offer a permanent solution to a pest problem. Consequently, biological control frequently is the method of choice in managing a pest species. Unfortunately, many pests are not suitable candidates for biological control and alternative methods need to be employed. Nevertheless, understanding the importance of natural enemies in controlling a pest species and recognizing potential impacts of management tactics on natural control are fundamental considerations in any pest management program.

A practical understanding of biological control requires that you can recognize basic groups of natural enemies (also called biological control agents). Additionally, you should be familiar with the three basic procedures of biological control.

Biological Control Agents

Insects are killed by a variety of organisms and these organisms fall into three classes: predators, parasites (including parasitoids), and pathogens. Organisms that feed on insects are said to be **entomophagous** (insect eating). Predators and parasites are entomophagous groups, whereas pathogens are disease-producing organisms.

Predators are free-living organisms that feed on other animals (prey). Predators kill

254

more than one prey item in order to reach maturity or through the course of adult life. Although some predators are very specific in their choice of prey species, generally predators have a wider host range than either parasites or pathogens. A number of vertebrate groups are predaceous on insects including birds, fish, reptiles (primarily lizards), amphibians (primarily toads and frogs), and mammals (primarily bats, shrews, moles, and small rodents). Birds are the most important vertebrate predators of insects, but birds only rarely have been used in biological control. One vertebrate species that is of some importance for biological control is the fish, *Gambusia affinis*, which feeds on mosquito larvae.

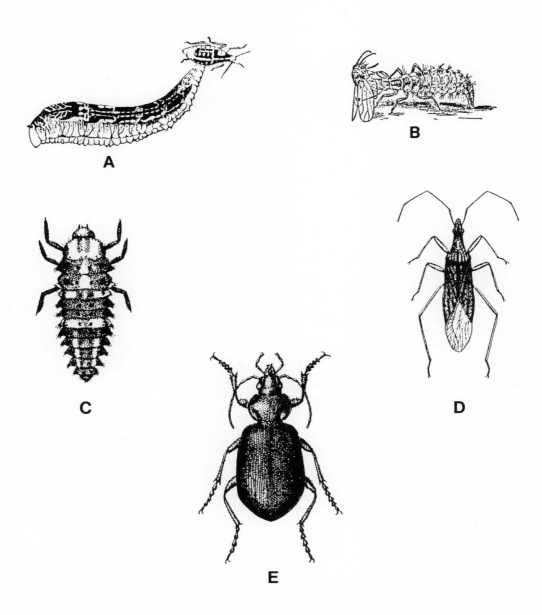

Figure 13.1   Examples of predaceous insects. A. Flower fly larva (Syrphidae) feeding on an aphid. B. Aphidlion (Chrysopidae) larva feeding on a plant louse. C. Ladybird beetle (Coccinellidae) larva. D. Damsel bug (Nabidae). E. Ground beetle (Carabidae). (A,C,D,E Courtesy Colorado State University; B Courtesy USDA)

However, both as natural and biological control agents, arthropods are much more important predators of insects than are vertebrates. Spiders, mites, and insects are the major groups of predaceous arthropods, and, ironically, insects themselves are the most significant single predaceous group. Table 13.1 summarizes the diversity among arthropod predators of insects. Some species are predaceous only as immatures or adults, although others are predaceous throughout their life cycle. Additionally, predators may feed on both immature and adult prey.

Table 13.1 Entomophagous arthropods. (Group = Class: Order; Families = number of families in order with at least one entomophagous species; these estimates will vary with different classification schemes.)

Group	Families	Group	Families
		PREDATORS	
Arachnida:		Insecta:	
Acari[1]	50	Coleoptera	51
Amblypygi	1	Dermaptera	2
Araneae	49	Diplura	1
Opiliones	7	Diptera	30
Palpigradi	1	Ephemeroptera	3
Pseudoscorpiones	20	Hemiptera	22
Ricinulei	1	Hymenoptera	33
Schizomida	1	Lepidoptera	11
Scorpoines	2	Mecoptera	3
Solifugae	2	Neuroptera	20
Uropygi	1	Odonata	16
Chilopoda:		Orthoptera	4
Geophilomorpha	5	Plecoptera	4
Lithobiomorpha	2	Psocoptera	3
Scolopendromorpha	2	Thysanoptera	3
Scutigeromorpha	1	Thysanura	1
Diplopoda:		Trichoptera	1
Cambalida	1		
Chordeumida	1		
		PARASITOIDS	
Insecta:			
Coleoptera	8		
Diptera	10		
Hymenoptera	46		
Lepidoptera	1		
Strepsiptera	4		

[1]Families with predaceous and parasitic spp.; (no parasitoids, but some parasitic mites routinely kill insect host).

Many insect orders have predaceous species; however, the most important orders in terms of diversity and usefulness for biological control are the Coleoptera, Diptera, Hemiptera, Hymenoptera, and Neuroptera. Some important predaceous families are the Carabidae (ground beetles) and Coccinellidae (ladybird beetles) in the Coleoptera, Syrphidae (flower flies) in the Diptera, Nabidae (damsel bugs) in the Hemiptera, Formicidae (ants) in the Hymenoptera, and the Chrysopidae (aphidlions or lacewings) in the Neuroptera.

Parasites are animals that live on or within another animal, the host. Arthropod parasites that kill their host are known as **parasitoids** and are of particular importance for biological control. Parasites include mites, insects, and nematodes, but all parasitoids are insects (although some nematode species may routinely kill their host; these are not called parasitoids and are sometimes referred to as pathogens). Parasitic mites may influence pest populations to a limited degree but not as dramatically as do parasitoids.

Parasitic nematodes can impose considerable mortality on certain insect pest populations, and three nematode families, Mermithidae, Neotylenchidae, and Steinernematidae, are especially important as insect parasites. Some efforts have been and are directed at using nematodes in biological control, but nematodes are less useful than other biological control agents.

In contrast, parasitoids are extremely useful and widespread biological control agents. Although only five insect orders include parasitoid species, a tremendous diversity and number of species occur (Table 13.1). The order Hymenoptera includes both the greatest number of parasitoid species as well as the most significant species in terms of effects on pest populations. The families Ichneumonidae and Braconidae contain many important biological control agents. All insects in these families are parasitic, and both contain a great many species. In fact, the Ichneumonidae is thought to contain more species than any other insect family and actually includes more species than many insect orders. Other important hymenopteran families with parasitoids are the Chalcidae, Encyrtidae, and Trichogrammatidae.

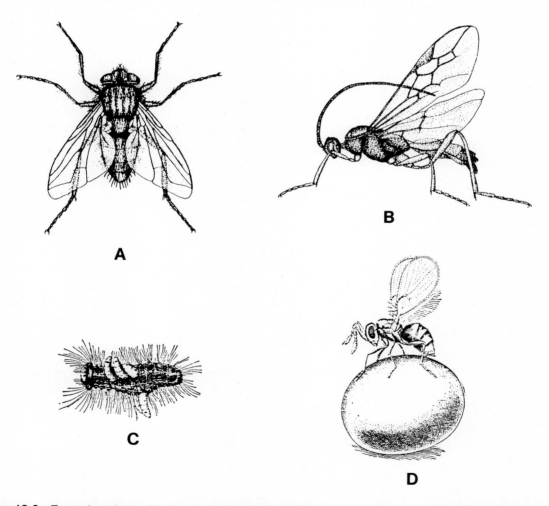

Figure 13.2   Examples of parasitic insects. A. Tachinid adult (Tachinidae), larvae are parasitic. B. Braconid adult (Braconidae). C. Braconid larvae (Braconidae) emerging from gypsy moth caterpillar. D. Trichogrammatid adult (Trichogrammatidae) ovipositing on a moth egg. (A,B Courtesy Colorado State University; C,D Courtesy USDA)

Pathogens are organisms that produce disease in a host, and insects are affected by various diseases just as are other types of animals. Many diseases are fatal, whereas others may reduce an insect's fitness or reproductive potential. The most important groups of insect pathogens are the bacteria, fungi, protozoa, rickettsia, and viruses. Of these, the bacteria and viruses have been of most use as biological control agents to date. For biological control, pathogens are used in **microbial insecticides**, biological preparations used in a manner similar to conventional chemical insecticides.

Basic Biological Control Procedures

The three fundamental procedures in biological control are introduction, augmentation, and conservation. In all of these the intent is to increase the effectiveness of natural enemies in reducing pest populations. Although some biological control activities can be undertaken by individual producers, others require involvement of many producers or governmental agencies.

Historically, the most effective biological control procedure has been introduction. **Introduction**, also called classical biological control, is the importation and establishment of natural enemies for a given pest. Introduction most often is used for introduced pests which may lack many natural enemies in their new habitat. When successful, introduction can provide control of a pest species so that the pest no longer occurs in economically important numbers. Introduction requires exploration and surveys for natural enemies in areas of the world where a pest originally occurred. Additionally, potential natural enemies must be quarantined, evaluated, and, if appropriate, released over a wide area. Consequently, introductions are conducted by governmental agencies rather than individual growers.

Augmentation, practices designed to increase numbers of natural enemies, is conducted by governmental agencies and individual growers. The rationale behind this approach is that natural enemies that do not sufficiently reduce pest populations may be effective for biological control if the natural enemy population is increased or augmented. Thus, augmentation involves the periodic release of natural enemies. One type of augmentation is **inoculative release**, in which relatively small numbers of natural enemies are released and are expected to increase and spread. Inoculative releases are similar to introductions, but the natural enemies associated with augmentation usually do not become established (for example, natural enemies may be killed during winter). The other type of augmentation practice is **inundative release**, in which large numbers of natural enemies are released. Inundative release requires mass rearing of natural enemies which presents a variety of problems. Use of microbial insecticides can be thought of as a valuable type of inundative release.

Conservation is any activity that protects and maintains existing natural enemy populations. Many pest management activities relate to conservation, as do other production practices. Practices in support of conservation include activities such as timing cultural operations, maintaining habitat and food sources for natural enemy populations, choice of insecticide, timing for insecticide applications, and reducing insecticide dosage.

GENETIC CONTROL

Genetic control is the management of insect pest populations through the manipulation of their genetic component or mechanisms of inheritance (also called **autocidal control** - self-killing). Although many manipulations of insect genetics have been researched, the only technique demonstrated to be practical is the sterile insect release method (also called the sterile-male technique).

The **sterile insect release method** involves replacing normal matings with infertile matings by flooding the population with sterile males, who mate with fertile females. These matings between sterile males and fertile females produce non-viable eggs, and therefore the pest population will decline. In the extreme, it may be possible to completely eliminate a pest from an area. Insects usually are sterilized by irradiation, but the sterilization technique must not impair insect fitness or mating behavior.

Table 13.2   Results of different levels of sterile insect releases for populations of a hypothetical pest (having a 1:1 sex ratio and 5-fold increase per generation).

| | MALES | | FEMALES | | |
Generation	Sterile Released	Ratio of Sterile:Fertile	With Release	Fertile-Mated	Without Release
			Low Release Rate		
1	2,000,000	2.00:1	1,000,000	333,333	1,000,000
2	2,000,000	0.60:1	1,666,665	1,041,666	5,000,000
3	2,000,000	0.38:1	5,208,328	3,774,151	25,000,000
4	2,000,000	0.11:1	18,870,754	17,000,679	125,000,000
5	2,000,000	0.02:1	85,003,395	83,336,662	625,000,000
			Moderate Release Rate		
1	4,000,000	4:1	1,000,000	200,000	1,000,000
2	4,000,000	4:1	1,000,000	200,000	5,000,000
3	4,000,000	4:1	1,000,000	200,000	25,000,000
4	4,000,000	4:1	1,000,000	200,000	125,000,000
5	4,000,000	4:1	1,000,000	200,000	625,000,000
			High Release Rate		
1	8,000,000	8:1	1,000,000	111,111	1,000,000
2	8,000,000	14:1	555,555	36,075	5,000,000
3	8,000,000	44:1	180,375	4,067	25,000,000
4	8,000,000	393:1	20,335	52	125,000,000
5	8,000,000	30,769:1	260	0	625,000,000

To be successful, a sterile insect release program must produce greater numbers of sterile males than fertile males, otherwise enough fertile matings will occur to keep the pest population high. Consequently, sterile insect releases involve the release of tens and hundreds of millions of sterile males for an extended period. As a sterile insect release program becomes successful, the proportion of sterile males to fertile males will increase, resulting in greater reductions in the pest population.

Table 13.2 illustrates how sterile insect releases can reduce a pest population over several generations. As illustrated in this hypothetical example, the proportion of sterile males to fertile males is extremely important in determining the influence of sterile male releases on pest populations. If too few males are released (e.g., the 2,000,000 release in Table 13.2), the pest population will continue to increase, or may become stable (e.g., the 4,000,000 release in Table 13.2). If sufficient sterile males are released (e.g., the 8,000,000 release in Table 13.2), the pest population will decline, and complete eradication of the pest population is theoretically possible. Because such large numbers of sterile insects must be released and distributed over a large area, sterile insect release programs are extremely expensive and must be undertaken by governmental agencies.

The sterile insect release method is most useful for eliminating or preventing an introduced pest population from establishing itself in a new area. For example, sterile insect release has been used in eradicating introductions of the Mediterranean fruit fly in California. Another important situation where sterile insect release has proved useful is in the eradication of isolated populations of pests, such as on islands or restricted habitats.

Many requirements must be met to use sterile insect release against an insect pest species. At the outset, substantial biological information on the pest species must be available. It must be possible to rear the pest insect in large numbers (greater than the target population in the field). Additionally, the sterilization procedure cannot significantly alter aspects of pest fitness or behavior, especially mating behavior. Releases of sterile insects must be conducted to allow adequate mixing of sterile and natural pest populations. Moreover, the released sterile insects cannot be pests themselves. A final important practical requirement is that because of the expense and politics associated with the use of sterile insect release programs, target pest species must be of great economic importance.

The classic example of the use of the sterile insect release method to eradicate an important insect pest was the elimination of the screwworm, *Cochliomyia hominivorax* (Fig. 13.3), on the island of Curacao near Venezuela in 1954. Subsequently, screwworm eradication was undertaken and achieved in the southeastern US, and more recently, screwworms have been eliminated from the western US and most of Mexico.

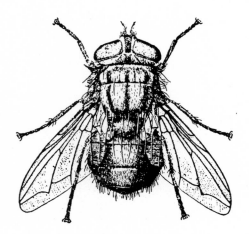

Figure 13.3 Screwworm adult. Screwworm larvae feed in open wounds of cattle. (Courtesy USDA)

OBJECTIVES

1) Know basic features of life history and how to recognize the most important insect predators including: ants, aphidlions (lacewings), damsel bugs, flower flies, ground beetles, and ladybird beetles.

2) Know basic features of life history and how to recognize parasitoids in the following families: Braconidae, Chalcidae, Ichneumonidae, and Trichogrammatidae.

3) Know the three basic procedures for biological control.

4) Understand the principles and basic requirements for a sterile insect release program.

5) Additional objectives (at instructor's discretion):

OPTIONAL DISPLAYS AND EXERCISES

1) Select a pest species and learn its associated natural enemies and their life cycles.

2) Examine displayed information on microbial insecticides.

3) Identify a pest species in your area that might be an appropriate candidate for genetic control. Determine the advantages and potential problems in employing sterile insect release for this pest.

TERMS

augmentation   genetic control   natural control   pathogens
autocidal control   inoculative release   natural enemies   predators
biological control   introduction   parasites   sterile insect release
conservation   inundative release   parasitoids   method
entomophagous   microbial insecticides

DISCUSSION AND STUDY QUESTIONS

1) Why is introduction likely to be less successful with native pests than with exotic (introduced) pests?

2) What are some advantages and disadvantages in using predators for biological control?

3) How are production systems (e.g. agronomic, horticultural, medical, urban) different in their suitability for biological control?

4) What are some problems with using augmentation as a management tactic?

5) Why isn't genetic control employed more frequently as a management approach? How do the principles of genetic control mesh with principles of IPM?

BIBLIOGRAPHY

Pedigo, L. P. 1989. Entomology and pest management. Macmillan Pub. Co., New York, NY
- Chapter 9. Natural enemies of insects and Chapter 14. Sterile-insect and other genetic tactics

Biological Control

Clausen, C. P. 1972. Entomophagous insects. Hafner Pub. Co., New York, NY
- an extremely informative book with substantial information on the various groups of entomophagous insects. Unfortunately out of print.

Coppel, H. C., and J. W. Mertins. 1977. Biological insect pest suppression. Springer-Verlag, New York, NY
- an overview of biological control, which is less comprehensive but more readable than many of the multi-authored texts. Includes a number of examples and a glossary.

DeBach, P., ed. 1964. Biological control of insect pests and weeds. Chapman and Hall, Ltd., London
Huffaker, C. B., and P. S. Messenger, eds. 1976. Theory and practice of biological control. Academic Press, Inc., New York, NY
- both provide good discussions on many aspects of biological control. Treatment of specific topics is somewhat uneven (as is the case with many multiauthored works), especially in Huffaker and Messenger. Nevertheless, both texts provide substantial information on the principles and practice of biological control, as well as numerous examples.

Genetic Control

Knipling, E. F. 1979. The basic principles of insect population suppression and management. USDA Agriculture Handbook No. 512
- provides a summary and tables regarding genetic control, particularly sterile insect release. Not particularly readable, but a useful reference nevertheless.

14 ECOLOGICAL MANAGEMENT AND HOST PLANT RESISTANCE

Many aspects of the crop environment and characteristics of crop plants may influence insect pest abundance and impact. The purposeful manipulation of the crop environment to reduce rates of pest increase or damage is termed **ecological management** or **cultural control**. Ecological management tactics fall into four main categories: (1) procedures aimed at reducing the average favorability of the ecosystem for pests, (2) procedures designed to disrupt the continuity of resources needed by pests, (3) procedures that attempt to divert pest populations from the crop, and (4) procedures aimed at reducing the impact of insect pest injury.

Like the cropping environment, crop plants may be modified to discourage pest problems. By using selective breeding and genetic engineering, plant breeders strive to develop plant varieties with qualities that make them undesirable or unsuitable for insect pests. We define any inherited characteristics of host plants that lessen or prevent attacks by pests as **host plant resistance**. Both ecological management and host plant resistance offer advantages over other management tactics: they can reduce our dependence on insecticidal management, and they can be economically sound and environmentally safe.

ECOLOGICAL MANAGEMENT

Reducing Ecosystem Favorability

Various procedures aim to reduce the favorability of pest ecosystems by lessening the average availability or suitability of food, shelter, or habitable space. For instance, **crop residue removal and destruction** (or **sanitation**) serves to eliminate materials in which pests may breed, overwinter, or congregate; this can diminish the size or impact of future populations. One important disadvantage of this procedure is that removal of plant residues, especially in row crops, encourages soil erosion and decreases the amount of organic material returned to the soil. Also, residue removal and destruction generally are effective only if practiced over a large geographical area. Otherwise, a neighboring grower's residues can easily serve as a source of pests which may invade other growers' crops.

Some pests can be managed through **destruction or modification of alternate hosts and habitats**. Many pests use or require resources that are supplied by hosts or habitats that lie outside of the crop. Elimination of alternate food sources (e.g., specific weeds) or alternate habitats (e.g., trash or plant cover in which overwintering occurs) can reduce pest abundance and impact. These tactics are useful when alternate requisites are manageable, that is, when only a small number of alternate hosts or habitats are present and available for removal or replacement. As with crop residue removal and destruction, alternate host and habitat management is most effective when employed over large geographical areas.

Tillage techniques sometimes may be used to change the physical environment of soil-inhabiting insects and, thus, reduce their numbers or impact. By timing normal tilling operations to correspond to soil insect pests' pupation or dormancy periods, the grower can sometimes expose the pests to conditions (e.g., weather, predators) which may impede survival. Of course, tillage for this purpose is not always feasible; it may not be economically sensible if fuel costs are high, or environmentally sound if it enhances soil erosion.

In some urban and stored product pest ecosystems, the **manipulation of environmental temperatures and humidities** is a feasible method of reducing the environmental favorability for many types of pests. Storing grains and other products under conditions unsuitable for pest growth and reproduction is a simple means of reducing the impact of such pests. Many types of temperature and humidity adjustments can be made without diminishing stored product quality, but others can harm stored goods. Consequently, you must always be mindful of product quality before making adjustments in storage conditions.

Irrigation and water management practices may also be exploited as means of limiting pest ecosystem suitability. Periodic flooding of fallow land to control soil pests, sprinkler irrigation for suppressing certain foliage-feeding pests, and water level management for altering mosquito breeding sites are some of the more common water management practices used (Fig. 14.1). Water management or irrigation generally is not feasible nor economically sound when used principally and solely for pest management; however, when these techniques fit comfortably into the normal agronomic-management scheme, then it may be wise and expedient to exploit their pest management utility.

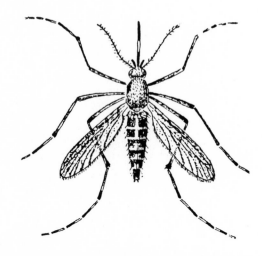

Figure 14.1 Salt-marsh mosquito. Various mosquito species have been controlled through water-management strategies. (Courtesy USDA)

Disrupting Ecosystem Continuity

Within the limits of good agronomic practices, we sometimes can **interrupt the supply of pest insect resources or requisites**. Such an interruption can be induced by manipulating the spacing of crop plants or crops, or the timing of planting and other operations. Certain **spacings of plants** within the crop may have adverse effects on insect pests. For example, denser planting sometimes results in higher humidities within the crop canopy and this may increase the occurrence or virulence of insect pathogens. Likewise, the **location of the crop** with respect to other crops and various non-cropping ecosystems is important and may be of utility in managing insect pests. It is usually wise to locate botanically-similar crops away from one another to avoid pest movement between crops.

Resource continuity also may be disrupted if we can create a gap, in time, in the insect food source. **Crop rotation,** or alternating the type of crop planted at a location, is one of the simplest and most important methods of providing discontinuity in a pest's requisites. Rotation schemes work best when the target pest has a narrow host range, lays eggs before the new crop is planted, and has a relatively immobile feeding stage. Crop rotation is a crucial technique for managing many plant diseases.

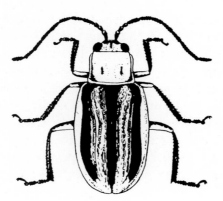

Figure 14.2   Western corn rootworm. Crop rotation is an effective and commonly used management tactic for this pest. (Courtesy Colorado Agricultural Experiment Station)

Modifying planting dates is another simple and widely-used method of achieving asynchrony between pests and crops. For example, by delaying planting until after a pest has laid the eggs that hatch to the damaging stage, infestation of the crop may be avoided. This technique works best for managing species of pests that oviposit simultaneously and that have a single generation per year. The timing of other cultural activities, such as the harvesting or cutting of crops and the dehorning and castration of livestock sometimes can be planned such that the host plants or animals are rendered unavailable or unsuitable at the time that the pests are active.

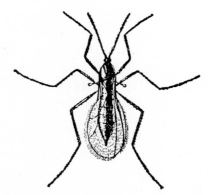

Figure 14.3   Hessian fly. Modifying planting date is an effective and commonly used management tactic for this pest. (Courtesy USDA)

Diverting Pests from Crops

Sometimes it is not practical, agronomically feasible, or economically realistic to modify the crop or nearby environments for the purpose of managing pests. This is especially true for insect pests with high rates of dispersal or with migratory capabilities. It is possible, however, to take advantage of some insects' dispersal patterns, or their preferences for certain hosts over others, by using certain diversion techniques such as trap cropping or strip harvesting.

A **trap crop** may consist of a different species or variety other than the crop you wish to protect, or it may be the same species or variety planted at a different time. Whatever the case, the trap crop should be more attractive to the pest than the protected crop; this compels pests to move into the trap crop and away from the protected crop. Pests then may develop undisturbed in the trap or may be killed in place with an insecticide.

Strip harvesting is similar to trap cropping except that a trap is created within the main crop when different areas, or strips, of it are harvested at different times. This encourages insect pests to stay within the stripped crop and discourages them from moving into adjacent, and often, more valuable crops. For example, when alfalfa and cotton are grown side by side, the alfalfa may be harvested in strips. This keeps insect pests in the alfalfa and thus prevents a mass exodus to the more valuable cotton crop.

Some mechanical devices, such a blacklight traps or pheromone-baited sticky traps, may aid in diverting certain insect pests away from a crop or commodity. Such traps are more frequently employed in pest detection and population sampling, but occasionally they are used for controlling insects by diverting and then physically removing them from the crop or commodity. Traps are most practical for management of flying insects in small, enclosed areas such as warehouses or livestock-confinement buildings. However, trap utility tends to be limited by their initial expense and their maintenance and replacement costs.

Reducing the Impact of Pest Injury

In a well-conceived pest management program, it is important to realize that the presence of some insect pests is unavoidable, and injury to crops and commodities is inescapable. Sometimes, therefore, we need to concentrate principally on managing and minimizing the inevitable injuries and losses sustained by the crop, rather than focusing on eliminating or diverting the crop's adversaries. Cultural techniques for minimizing the impact of pest injury include tactics designed specifically for this purpose, such as altering harvest schedules, but also encompass any practices that contribute to improved plant or animal health. It is a salient but often overlooked fact that healthy, vigorously-growing crop plants and livestock more readily withstand, tolerate, or recover from attack by insect pests than do weaker, ill-managed plants and animals.

Altering harvest schedules is one of the most frequently-used methods of reducing pest injury impact. Sometimes early harvesting allows the crop to escape the heaviest damage by pests. This technique often is employed in forage crops where other management tactics, such as insecticide application, are not economically feasible. Although early harvesting in grain crops to reduce the impact of insect pest injury also may alleviate certain pest problems, it is seldom practiced because early-harvested grain often has a high moisture content and must undergo expensive drying prior to storage.

HOST PLANT RESISTANCE

Of all the ecological management practices used to minimize injury and losses, the planting of crop plant varieties genetically resistant to insect damage is the most important tactic in terms of both the acreage and the number of crops in which it is employed. Resistance is a valuable and flexible tactic because it usually is compatible with other cultural techniques, insecticidal management, and biological control.

The Genetic Nature of Host Plant Resistance

The concept of using host resistance to our advantage comes from the knowledge that most plants and animals are resistant to most potential insect attackers. Various physiological, morphological, or behavioral characteristics inherited by plants and animals serve as a core of defense against species that would otherwise attack and consume them. These defenses are the result of natural selection; resistance traits are preadaptive characteristics of plants and animals. Organisms possessing genes conferring these preadaptive traits are better able to withstand the selective pressures of pests, thus improving chances for survival and successful reproduction. It is the job of the plant breeder to identify and isolate lines of plants possessing genes that convey resistance traits and to then transfer the resistance genes into lines with desirable agronomic characteristics. Presently, breeding for insect resistance in crop plants is far more advanced than it is in domesticated animals. Because of its significance in plant pest management, and lesser importance in animal production, only plant resistance will be discussed henceforth.

Mechanisms of Plant Resistance

Resistant plants may reduce the effects of insect pest attack in different ways. Typically, the modes or mechanisms of resistance are classified as nonpreference, antibiosis, or tolerance.

Nonpreference or **antixenosis** refers to plant characteristics that lead insects away from a particular host. These characteristics do not actually cause direct harm to the insect pest but do involve participation of both the plant and the pest. Nonpreference may be expressed in a cultivar through either allelochemic or morphological characteristics.

Allelochemic nonpreference is exhibited by plants or cultivars which possess or lack particular chemicals that either stimulate and attract, or inhibit and deter, insects and subsequent insect behaviors (e.g., feeding, oviposition). For instance, a cultivar lacking a chemical, called a **kairomone**, which is needed to stimulate insect sensory receptors and initiate subsequent insect behaviors such as feeding, would be considered nonpreferred, while a cultivar possessing the kairomone would be the preferred host. Similarly, a cultivar possessing a chemical, known as an **allomone**, which interferes with host recognition or selection (e.g., by repelling the insect or suppressing feeding responses) would be nonpreferred compared to a cultivar lacking the allomone. Allelochemic nonpreference is most effective against **monophagous** insects, or those with narrow host ranges. This is because monophagous species tend to have more sensitive sensory receptors than do **polyphagous** species, those with broad host ranges. The sensitive receptors of the monophagous species are what permit the insects to distinguish between an acceptable and unacceptable host. Only hosts possessing the proper allelochemics, needed to stimulate sensory receptors and incite subsequent feeding, will be accepted by many monophagous species. In fact, nonpreference can be so strong that certain insects may actually starve to death in the absence of foods containing the proper allelochemics, even though the other available foods would cause the insects no harm.

Morphological nonpreference is the result of plant structural characteristics which disrupt normal insect behaviors by physical means. Excess or reduced pubescence, coloration changes, exudation alterations, or modifications in tissue characteristics all may serve to alter or inhibit normal insect responses toward a plant or plant part and thus may render a plant nonpreferred. Forms of morphological nonpreference that impair feeding behavior are very important as a first line of defense against pests and tend to provide longer-lasting effectiveness than do allelochemically-based forms of nonpreference.

Another mechanism of plant resistance is termed **antibiosis**. Antibiosis is the tendency of a plant to injure or destroy insect life. Antibiosis may be the result of either biophysical or biochemical factors and, like nonpreference, involves participation by both the plant and the insect. Symptoms of insects affected by antibiosis include: death of young immatures, reduced growth rate and increased duration of the period between molts, increased pupal mortality, smaller adults, decreased fecundity, shortened adult lifespan, morphological malformations, lack of formation of proper food reserves, restlessness, and other abnormal behaviors.

Biophysical antibiosis results from structural variations in the plant which cause harm to insects. For instance, sticky or pointed hairs and spines or gummy exudates can ensnare an insect on the plant's surface, and tough, solid stems may crush or suffocate a burrowing insect or significantly slow its progress toward suitable food.

Biochemical antibiosis usually results from insect consumption of plant-produced chemicals which function as toxins or metabolic inhibitors. It also may be the result of nutritional deficiencies that result when an insect consumes a plant lacking certain nutrients or containing an imbalance of particular dietary components (e.g., sugars, amino acids).

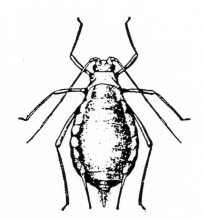

Figure 14.4   Pea aphid. Some alfalfa cultivars have resistance to pea aphid through biochemical antibiosis. (Courtesy USDA)

The plant resistance mechanism known as **tolerance** differs from nonpreference and antibiosis in that it depends only on a plant response; insect responses are not involved. Tolerance is defined as the ability of a plant or cultivar to give satisfactory yields in spite of injury levels that would debilitate nonresistant plants. This is the least dramatic of the resistance mechanisms, and some scientists do not consider it a form of resistance.

There are many components of tolerance including: general plant vigor, hybrid vigor, compensatory growth (by individual plants or the plant population), wound healing and regrowth, improved tissue strength, and changes in photosynthetic partitioning. Disadvantages of tolerance include that it is more subject to modification by environmental factors than are nonpreference or antibiosis and that it does not cause a reduction in the size of pest populations. The primary advantage of tolerance is its permanence; tolerance places no selective pressure on an insect pest population. Therefore, no variants of pests are likely to develop that can overcome the resistance.

Advantages of Host Plant Resistance

The employment of host plant resistance as a pest management tactic is limited primarily by our ability to identify and locate sources of resistance genes and our success in transferring them into plant lines with acceptable agronomic qualities. In spite of these limitations, the advantages of plant resistance are great. First, resistant crop plants which exhibit the mechanisms of nonpreference or antibiosis have a specific, cumulative, and relatively persistent negative impact on pest populations. Additionally, resistance leaves no toxic residues in the environment and presents little or no danger to the grower or to nontarget organisms. Also, resistance may be employed at relatively low cost to the grower since the cost of the resistance is incorporated into seed prices. Finally, resistance may be used in harmony with virtually all other pest management tactics.

OBJECTIVES

1) Know and understand the four major types of ecological management tactics.

2) Understand the disadvantages and advantages of specific ecological management tactics.

3) Be aware of the genetic nature of host plant resistance and how this might affect or limit its employment.

4) Understand the primary mechanisms of plant resistance: nonpreference, antibiosis, and tolerance.

5) Additional objectives (at instructor's discretion):

OPTIONAL DISPLAYS AND EXERCISES

1) Using any resources available to you, determine what ecological management techniques are feasible or recommended for one or more of the following situations. What are the advantages, disadvantages, and limitations of these techniques?

    a. Hessian fly in wheat
    b. western corn rootworm in field corn
    c. stable flies in a cattle feedlot
    d. smaller European elm bark beetle on American elm
    e. potato leafhopper in alfalfa
    f. squash bugs on commercial vegetables

2) Plant several varieties of soybeans, alfalfa, or wheat. Cage the plants and then expose them to associated pest insect species (e.g., Mexican bean beetle, pea aphid, cereal leaf beetle). Evaluate each variety in terms of its resistance to insect attack. How do you results compare to published data? Which mechanism of resistance to you suspect to be responsible in the varieties exhibiting resistance?

3) Examine the traps and other mechanical devices on display that are sometimes used to divert and physically remove or destroy pest insects from a crop or commodity.

4) Additional displays and exercises (at instructor's discretion):

TERMS

allelochemic	crop rotation	kairomone	sanitation
nonpreference	cultural control	monophagous	strip harvesting
allomone	ecological	nonpreference	tolerance
antibiosis	management	polyphagous	trap cropping
antixenosis	host plant resistance		

DISCUSSION AND STUDY QUESTIONS

1) If you were conducting a research program to locate plant lines that possess genes to confer resistance to a pest insect, would you have a better chance for success if you were working with a monophagous or polyphagous insect? Explain.

2) Several lines of a crop plant were grown in field plots that were uniformly infested with root-feeding insect larvae. Later inspection of the root systems of the plants revealed that certain lines exhibited larger root systems, primarily due to greater branching and proliferation of

roots above the larval feeding sites. There were no differences between lines in terms of the number of larvae found feeding on the root systems. What mechanism of resistance caused the differences between lines? Explain.

3) Circular disks of leaf tissue were cut out of alfalfa leaves from several varieties. The disks were placed around the perimeter of a petri dish. Adult alfalfa weevils were released in the center of the dish, the lid was placed on the dish, and after one hour, the number of alfalfa weevils feeding on each leaf disk was recorded.   Some alfalfa varieties were consistently found to have fewer beetles feeding on them. What mechanism of resistance probably caused the differences between varieties? Explain.

4) Insect pests such as grasshoppers, stalk borers, and armyworms usually are problems only at the margins of field crops. Why do you suppose this is so? What ecological management techniques might be useful in dealing with these pests?

5) Speculate on how increasing the spacing between plantings might reduce the numbers or impact of a hypothetical insect pest.

BIBLIOGRAPHY

Pedigo, L. P. 1989. Entomology and pest management. Macmillan Pub. Co., New York, NY
- Chapter 10. Ecological management of the crop environment and Chapter 12. Plant resistance to insects.

Altieri, M. A. 1983. Agroecology: the scientific basis of alternative agriculture. Division of Biological Control, University of California, Berkeley, CA
- an analysis of novel agroecosystems and technologies, and sustainable agriculture. It contains an interesting chapter on the "agroecological basis for insect pest management" that focuses on vegetational diversity and its influence on insect populations.

Green, M. B. and Hedin, P. A., eds. 1986. Natural resistance of plants to pests: roles of allelochemicals. American Chemical Society, Wash. D.C.
Hedin, P. A., ed. 1982. Plant resistance to insects. American Chemical Society, Wash. D.C.
Hedin, P. A., ed. 1977. Host plant resistance to pests. American Chemical Society, Wash. D.C.
- these books are comprised of research papers by various workers examining the mechanisms, especially at the biochemical level, of plant resistance to pests. Although many of the papers are technical in nature, they give valuable insight into the complexities and intricacies of host plant resistance.

Martin, H. and P. Woodcock. 1983. The scientific principles of crop production. Edward Arnold Publ., Baltimore, MD
- a fairly complete analysis of a variety of pest management tactics, including plant resistance and various ecological management techniques.

Maxwell, F. G. and P. R. Jennings, eds. 1980. Breeding plants resistant to insects. John

Wiley & Sons, Inc., New York, NY
- a useful, multiauthor publication covering many aspects of plant resistance mechanisms and breeding for resistance in specific crops.

National Academy of Sciences. 1969. Insect pest management and control. Principles of Plant and Animal Pest Control. Vol. 3. U.S. National Academy of Sciences Publication 1695
- a survey of insect pest management tactics which includes fairly comprehensive, although somewhat dated, chapters on plant and animal resistance to insects, cultural control, and physical and mechanical control.

Painter, R. H. 1951. Insect resistance in crop plants. University Press of Kansas, Lawrence, KS
- a classic text on the subject of host plant resistance. It outlines the mechanisms of resistance and the factors that affect the expression and permanence of resistance. Additionally, it provides detailed discussions of resistance in a variety of crops

Panda, N. 1979. Principles of host-plant resistance to insect pests. Allan Held, Osmun & Co., New York, NY
- this text provides an excellent history of the development of resistance to pests in many crops and also gives coverage to host-selection theories and to the mechanisms and genetics of resistance.

Pimentel, D., ed. 1981. CRC handbook of pest management in agriculture. Vol. II. CRC Press, Inc., Boca Raton, FL
- while the primary focus of this volume is biological control of pests, there also is a small section on cultural management.

Russell, G. E. 1978. Plant breeding for pest and disease resistance. Butterworths, London
- a well-researched text with excellent sections covering general principles and methods of breeding for resistance and examples of successful implementation of host plant resistance.

15 REGULATORY MANAGEMENT

Preventing the entry and establishment of foreign, or **exotic**, insect pests to a region, and eradicating, containing, or suppressing pests established in limited areas, are the basic tenets of regulatory insect management. The importance of regulating the introduction and spread of exotic insect pests cannot be underemphasized. In the US today, many of the most damaging insect pests of field, orchard and garden crops, and of stored goods are of foreign origin and were first introduced with plant material, in the course of commerce, or in some instances, intentionally. A partial listing of these introduced species (Table 15.1 and Fig. 15.1) illustrates the importance and prevalence of imported pests. Had the entry of these insects been prevented or their spread hindered, the faces of agriculture, commerce, and pest management might be very different today.

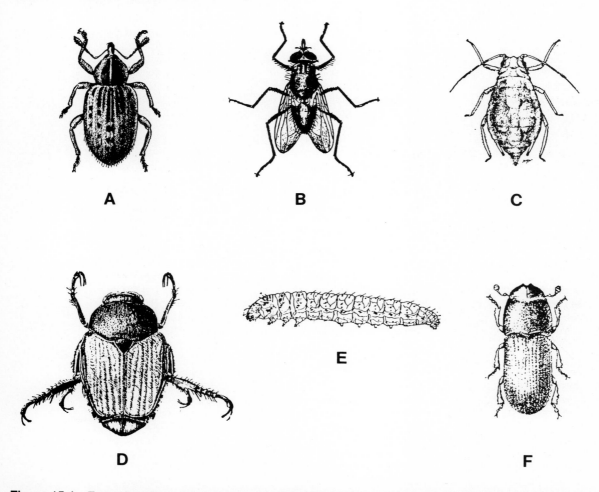

Figure 15.1 Examples of exotic insect pests now established in the U.S. A. Alfalfa weevil. B. Horn fly. C. Greenbug. D. Japanese beetle. E. Pink bollworm. F. Shothole borer. (A,B,C,D,E Courtesy USDA; F Courtesy Oklahoma State University)

273

Table 15.1 Some exotic insect pests now established in the US.

Common Name	Scientific Name	Order: Family
alfalfa weevil	*Hypera postica*	Coleoptera: Curculionidae
Angoumois grain moth	*Sitotroga cerealella*	Lepidoptera: Gelechiidae
Argentine ant	*Iridomyrmex humilis*	Hymenoptera: Formicidae
Asian cockroach	*Blattella asahinai*	Orthoptera: Blattellidae
Asian tiger mosquito	*Aedes albopictus*	Diptera: Culicidae
Asiatic garden beetle	*Maladera castanea*	Coleoptera: Scarabaeidae
asparagus beetle	*Crioceris asparagi*	Coleoptera: Chrysomelidae
blue alfalfa aphid	*Acyrthosiphon kondoi*	Homoptera: Aphididae
boll weevil	*Anthonomus grandis grandis*	Coleoptera: Curculionidae
browntail moth	*Euproctis chrysorrhoea*	Lepidoptera: Lymantriidae
cereal leaf beetle	*Oulema melanopus*	Coleoptera: Chrysomelidae
cigarette beetle	*Lasioderma serricorne*	Coleoptera: Anobiidae
citricola scale	*Coccus pseudomagnoliarum*	Homoptera: Coccidae
clover leaf weevil	*Hypera punctata*	Coleoptera: Curculionidae
clover seed weevil	*Tychius picirostris*	Coleoptera: Curculionidae
codling moth	*Cydia pomonella*	Lepidoptera: Tortricidae
cottonycushion scale	*Icerya purchasi*	Homoptera: Margarodidae
crazy ant	*Paratrechina longicornis*	Hymenoptera: Formicidae
diamondback moth	*Plutella xylostella*	Lepidoptera: Plutellidae
English grain aphid	*Macrosiphum avenae*	Homoptera: Aphididae
European chafer	*Rhizotrogus majalis*	Coleoptera: Scarabaeidae
European corn borer	*Ostrinia nubilalis*	Lepidoptera: Pyralidae
European earwig	*Forficula auricularia*	Dermaptera: Forficulidae
European pine sawfly	*Neodiprion sertifer*	Hymenoptera: Diprionidae
European pine shoot moth	*Rhyacionia buoliana*	Lepidoptera: Tortricidae
European red mite	*Panonychus ulmi*	Acari: Tetranychidae
face fly	*Musca autumnalis*	Diptera: Muscidae
Formosan subterranean termite	*Coptotermes formosanus*	Isoptera: Rhinotermitidae
greenbug	*Schizaphis graminum*	Homoptera: Aphididae
gypsy moth	*Lymantria dispar*	Lepidoptera: Lymantriidae
Hessian fly	*Mayetiola destructor*	Diptera: Cecidomyiidae
horn fly	*Haematobia irritans*	Diptera: Muscidae
imported cabbageworm	*Artogeia rapae*	Lepidoptera: Pieridae
imported currantworm	*Nematus ribesii*	Hymenoptera: Tenthredinidae
Indianmeal moth	*Plodia interpunctella*	Lepidoptera: Pyralidae
Japanese beetle	*Popillia japonica*	Coleoptera: Scarabaeidae
juniper webworm	*Dichomeris marginella*	Lepidoptera: Gelechiidae
Mediterranean flour moth	*Anagasta kuehniella*	Lepidoptera: Pyralidae
mimosa webworm	*Homadaula anisocentra*	Lepidoptera: Plutellidae
oriental beetle	*Anomala orientalis*	Coleoptera: Scarabaeidae
oriental fruit moth	*Grapholita molesta*	Lepidoptera: Tortricidae
pea aphid	*Acyrthosiphon pisum*	Homoptera: Aphididae
pea moth	*Cydia nigricana*	Lepidoptera: Tortricidae
pear psylla	*Psylla pyricola*	Homoptera: Psyllidae
pear sawfly	*Caliroa cerasi*	Hymenoptera: Tenthredinidae
pink bollworm	*Pectinophora gossypiella*	Lepidoptera: Gelechiidae
San Jose scale	*Quadraspidiotus perniciosus*	Homoptera: Diaspididae
shothole borer	*Scolytus rugulosus*	Coleoptera: Scolytidae
smaller European elm bark beetle	*Scolytus multistriatus*	Coleoptera: Scolytidae
sorghum midge	*Contarinia sorghicola*	Diptera: Cecidomyiidae
southwestern corn borer	*Diatraea grandiosella*	Lepidoptera: Pyralidae
spotted alfalfa aphid	*Therioaphis maculata*	Homoptera: Aphididae
spotted asparagus beetle	*Crioceris duodecimpunctata*	Coleoptera: Chrysomelidae
sugarcane borer	*Diatraea saccharalis*	Lepidoptera: Pyralidae
whitefringed beetles	*Graphognathus* spp.	Coleoptera: Curculionidae

HISTORICAL BACKGROUND

In the early days of agricultural trade, plants, animals, and their products were carried to and from different regions of the world with little or no regard for the pest-spreading potentials of the unregulated passage of goods. The first regulatory measure prohibiting the entry of insect-infested materials was instituted in Germany in 1873 to protect against entry of the grape phylloxera, *Daktulosphaira vitifoliae*, from North America. The first regulatory legislation in the US was passed in 1877, when four states enacted laws to protect their agricultural interests from particular pests. In 1905, the US Congress approved the **Insect Pest Act**, which provided the authority to regulate the entry and interstate movement of articles that might spread injurious insects. The **Plant Quarantine Act**, passed in 1912, authorized the Secretary of Agriculture to enforce regulations intended to protect the US agricultural economy by preventing the introduction of exotic insects and plant diseases. Unfortunately, by 1912 the US already had become a refuge for a variety of important insect immigrants, including the San Jose scale, gypsy moth, and boll weevil. Additional pertinent legislation included the Postal Terminal Inspection Act of 1915, amendments to the Plant Quarantine Act, the Mexican Border Act of 1942, the Export Certification Act of 1944, and several more recent state and national legislative actions. This legislation provides the authority to examine goods as they cross international boundaries into the US, or traverse regions within the US which are under quarantine for specific pests.

Today, APHIS, the Animal and Plant Health Inspection Service of the USDA, is largely responsible for the protection of American agriculture from exotic insects and plant diseases. The responsibilities of APHIS include: (1) preventing the entry of destructive pests into the US, (2) tracking the movement and range expansions of potential and threatening pests, (3) maintaining a pest survey and detection network, and (4) containing and eradicating exotic pests if they should enter the US.

QUARANTINE

Quarantine actions include all activities that serve to exclude potential pests from a region and restrict further spread of pests already present. The primary objective of all quarantine actions is the protection of economy and welfare against assault by exotic pests. Quarantines applied to goods at ports of entry are the first line of defense against the introduction of new pests. Such quarantines may apply to goods incoming with travelers and their accouterments, on commercial transportation, on privately-owned transportation, and in trade vessels and their cargo. These quarantines prohibit the importations of materials that may carry insects (e.g., raw fruits, other plant materials) or they require that such goods be inspected and treated, usually with fumigation, before being released.

Quarantines devised for preventing entry of insects from foreign countries (in particular, from those countries that do not share a border with the country instituting the quarantine) generally are more successful than are quarantines intended to restrict movement of insects within a nation. Intranational or interstate quarantines do not operate as effectively as international quarantines because large natural barriers (e.g., oceans) are seldom present to inhibit the spread or migration of pests. Also, central inspection points, such as seaports and airports, do not exist, and with the complicated array of highways and railways that traverse and link a nation, inspection of all goods entering and leaving an area usually is neither practical nor economical. The probability of success of an interstate quarantine is increased when natural barriers (e.g., a mountain range) are present, when access to and from the quarantined area is reduced, and when the quarantine is supported by other management practices aimed at reducing the pest population.

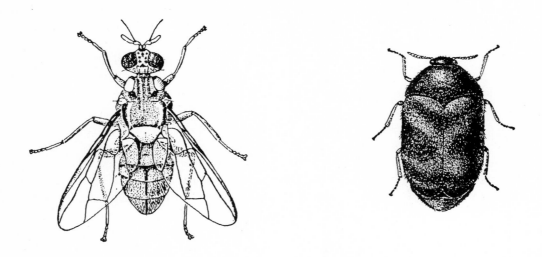

Figure 15.2 Examples of insect pests not yet established in the US and under federal quarantine. A. Oriental fruit fly. B. Khapra beetle. (Courtesy USDA)

If a pest crosses the first lines of defense and enters a new region, a quarantine may be imposed to prevent a moderate and defined infestation from spreading across additional available and ecologically suitable regions. When a pest is confined to a limited portion of its potential ecological range, a quarantine action is usually restricted to the source of the infestation. However, when the pest is widely distributed, quarantine actions are most practically applied to the periphery of the infestation or at the destinations to which goods from the infested area are destined.

Before invoking a quarantine, it is important to determine that a pest is a potential economic threat. Quarantines aimed at specific pests should not be based on the pests' importance in the regions of origin. Because exotic pests usually enter a new area unaccompanied by their associated predators, parasites, and pathogens, they may be of much more serious consequence in their new home than they were in their native habitat. This is why domestic quarantines are frequently initiated against pests that do not cause appreciable problems in other countries.

After a quarantine is adopted, continuous surveillance within the protected areas is important to assess the effectiveness of the quarantine. The presence of pests outside of the quarantined area indicates weaknesses or deficiencies in inspection programs or suggests increased population growth or pest dispersal.

ERADICATION, CONTAINMENT, AND SUPPRESSION

There are three primary types of publicly organized, widespread regulatory programs: eradication, containment, and suppression. These programs may utilize a variety of pest management tactics including insecticide application, biological control, and ecological management.

Eradication programs purport to totally eliminate the target pest from a defined geographical area. Such programs usually are applied only against pests that have recently immigrated to an area and have not yet spread over their potentially acceptable range. Successes in pest eradication include programs employed against the screwworm, *Cochliomyia hominivorax*, in the southeastern US in the late 1950's and the Mediterranean fruit fly, *Ceratitis capitata* (Fig. 15.3), in the Santa Clara Valley of California in the early 1980's.

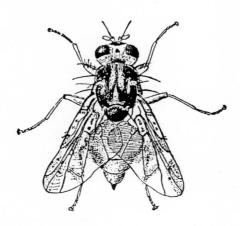

Figure 15.3   Mediterranean fruit fly. (Courtesy USDA)

Pest eradication depends on constant surveillance to detect new invasions of potential pests, and on timely, widespread corrective reaction to pests upon their detection. Without prompt response carried out over broad areas, pest populations may spread beyond the point where eradication is feasible or possible. Eradication of pests sometimes also must be repeated (i.e., it is never a final solution to a particular pest problem). For instance, while the Mediterranean fruit fly was successfully eradicated in California in the 1980's, it was also previously eradicated in Florida in 1929-1930, 1956, 1962, and 1963 and in Texas in 1966.

Containment of pests that have not achieved distribution across their potential range is conducted when eradication is not possible, due to the lack of eradicative treatments or to the pest becoming firmly and irrevocably established. Containment is any effort to limit the spread of insects into a larger area. Usually, there is no effort to reduce or limit populations within the infested area. Instead, control measures are confined to portions of infested areas from which pests are likely to be either artificially or naturally spread. Containment programs have long been employed to curb the spread of the gypsy moth within and from the US and have been suggested to slow the northward movement of the Africanized honey bee, *Apis mellifera scuttellata*, from Central America.

Suppression involves any program employed when sudden, periodic pest outbreaks occur over large areas and cannot be suitably dealt with by individual citizens. Suppression programs usually are publicly supported and are instituted by state and federal agencies in cooperation with private growers and grower organizations. Suppression programs occasionally are employed against grasshopper outbreaks in the western and central US.

THE NEED FOR REGULATORY MANAGEMENT:
SOME CASE HISTORIES

Despite quarantines and other regulatory efforts, pests all too frequently gain entry and establish themselves in new areas. The following descriptions of successful exotic introductions that have wreaked havoc in their new homes illustrate the severity of the threats posed by foreign pests and indicate the need for stringent quarantines. Following are examples of pests that have gained entry on plant materials, in other goods, and through intentional, purposeful importation. These examples demonstrate how devastating and irrepressible introduced pests may be.

Smaller European Elm Bark Beetle

Insect pests frequently have been imported along with plant or animal materials. For instance, the Hessian fly, *Mayetiola destructor*, probably was imported in straw bedding used by Hessian soldiers in the Revolutionary War; the horn fly, *Haematobia irritans*, was brought in with cattle shipments in the 1880's; and the smaller European elm bark beetle, *Scolytus multistriatus*, was introduced in 1930 on Carpathian elm logs used for furniture veneer. The smaller European elm bark beetle was reported in the US as early as 1909, but the 1930 shipment was notable in that those insects did not arrive alone; they carried a virulent and aggressive strain of the destructive parasitic fungus *Ophiostoma ulmi*, which causes Dutch elm disease. The subsequent epidemic nearly decimated the American elm tree population and forever changed the appearance of American boulevards, parks, and farmsteads.

Early eradication efforts aimed at the smaller European elm bark beetle proved futile. One of the many reasons for eradication's failure was that a native alternate vector for the fungal pathogen was already present in the US. The native elm bark beetle, *Hylurgopinus opaculus*, proved as efficient a vector of the causal fungus as the exotic bark beetle. Both beetles can carry the fungus from diseased trees and from dead elms or elm logs in which it is growing to healthy trees. Additionally, the disease can be spread from tree to tree via naturally occurring root grafts. An integrated pest management approach that employed a medley of tactics for managing the tree, insects, and the fungus was effective in reducing losses to Dutch elm disease in specific instances, but overall, did little to stop the epidemic from destroying the vast majority of American elms in the US.

Asian Tiger Mosquito

The recent introduction of *Aedes albopictus*, the Asian tiger mosquito, to the US illustrates that quarantine of food and food products alone is not adequate for protecting human wellbeing. *Aedes albopictus* was discovered in Harris County, Texas in 1985; this was the first time that this species was known to have become established in this hemisphere. It is believed that the insect was, and is, being introduced into the US in the water collected in imported, used automobile and truck tires. Tires that collect rainwater constitute a suitable breeding site for the mosquito, which normally lays eggs in tree holes and other water-holding containers of natural or human origin. Imports of used tires have increased significantly since 1982; Japan,

Taiwan, Korea, India, and the Philippines export large numbers of tires to the US and smaller numbers to Central and South America. Interstate commerce, especially trade in used tires, has served to spread the insect to several areas in the US. There is little doubt that the Asian tiger mosquito has the capability to spread throughout the continental US. The insect is cold-hardy and can survive in even the most northern states. Additionally, *Aedes albopictus* may present serious threats to the health and comfort of the American people. It is a known epidemic vector of dengue and dengue hemorrhagic fever in southeast Asia and is a highly efficient vector of La Cross encephalitis virus. The mosquito also is very annoying in that it exhibits very aggressive biting habits and may bite at all hours of the day.

The outlook regarding *Aedes albopictus* is rather gloomy despite the fact that the Department of Health and Human Service's Centers for Disease Control began studying the insect's biology, distribution, and geographic spread soon after its discovery. Unfortunately, the insect will probably continue to spread and thrive unless extremely vigorous eradication and containment programs are instituted soon. This mosquito is a vigorous breeder with a broad ecological range. In addition, both rural and urban environments can provide all the requisites (especially, the water-holding containers) needed for its survival and population expansion. It appears to be a likely candidate for continued spread and establishment across the US.

Gypsy Moth

Not all exotic pests have been introduced to the US by accident. The gypsy moth, *Lymantria dispar*, provides an example of an intentionally imported pest. Because cotton was unavailable during and shortly after the Civil War, the Union states were in need of a new fiber source. In 1869, M. Leopald Trouvelet, an American naturalist, attempted to rectify the fiber shortage by developing an American commercial silk industry. He imported a number of gypsy moth egg masses from southern France to Massachusetts. He had hoped to cross the gypsy moth with the silkworm moth to produce a hardy race of silk-producing insects. Unfortunately, during the course of his experiments some of the insects escaped and immediately began to gain a secure and permanent foothold in North American woodlands.

By 1880 the gypsy moth had infested 400 square miles around its release site, and by 1905, 2200 square miles of woodlands were infested. As the insect spread to neighboring states, Congress appropriated federal funds to prevent further spread. However, by 1922 the gypsy moth was distributed throughout New England. Since 1869, the natural spread of the gypsy moth has progressed at the rate of eight to ten miles per year. Its range covered over 200,000 square miles of the US and Canada by 1974, and over $110,000,000 had been spent by the federal government and millions more by state governments to combat the pest. Numerous efforts aimed at containing the species remain in effect today, and vigorous programs to monitor the insect's progress are carried out throughout the US. Such programs include an annual, national gypsy moth detection program. This program represents a cooperative effort involving participation by USDA-APHIS, US Department of Interior, Corps of Engineers, and various state and local government agencies. The program, which utilizes sex-pheromone-baited traps, has been useful in detecting isolated gypsy moth infestations toward which subsequent appropriate and expedient eradication efforts may be directed. While the program probably will not, on its own, solve the gypsy moth problem or ultimately prevent its spread across the forested regions of the US, it will, in combination with quarantine efforts, at least hinder the insect's progress and contribute toward prevention or limitation of isolated outbreaks.

OBJECTIVES

1) Understand the principles of quarantine, eradication, containment, and suppression.

2) Realize the abundance and importance of exotic pests in the US.

3) Be aware of the complexities and difficulties of regulatory management and how these may affect or limit its success or utility.

4) Additional objectives (at instructor's discretion):

OPTIONAL DISPLAYS AND EXERCISES

1) Examine the displayed examples of exotic pests now established in the US.

2) Using any resources available to you, prepare a brief report on the history, current pest status, and future outlook on one or more of the pests listed in Table 15.1.

3) Additional displays and exercises (at instructor's discretion):

TERMS

APHIS	eradication	Insect Pest Act	quarantine
containment	exotic	Plant Quarantine Act	suppression

DISCUSSION AND STUDY QUESTIONS

1) Not all exotic pests reach the US via travelers, imported goods, or transportation vehicles. Speculate on other ways in which exotic insects may enter a new region.

2) The southwestern desert once was an effective natural barrier to insect pests entering the US from Mexico. The boll weevil and numerous other pests were able to overcome this barrier. Why? What role did humans play?

3) With certain containment programs, pest management efforts are aimed at reducing popula-

tions within the infested area, with the hope that research breakthroughs and scientific discoveries will make eradication feasible and practical at a later date. What are the pros and cons of such an approach?

4) Some exotic pests that enter a new region thrive and prosper in their new habitat; others, most fortunately, are unable to survive and reproduce. What qualities make a pest a good candidate for successful immigration, dispersal, and establishment in a new location? Which types of pests are less likely to gain a foothold in a new region?

BIBLIOGRAPHY

Agricultural Research Service, USDA 1978. A handbook of pests, diseases, and weeds of quarantine significance. Amerind Publ. Co., New Delhi
- translated from Russian, this handbook provides an abundance of information on pests, diseases, and weeds that threaten the USSR. The pest reference information is thorough and includes complete, though slightly dated, descriptions of the geographic distribution and biology of pests that pose threats to the USSR., and in many cases, to the US as well.

Bombin, L. M. 1984. Plant protection legislation. Food and Agricultural Organization of the United Nations, Rome
- a study of plant protection legislation of 14 countries. It lists legal difficulties encountered in the various national laws and indicates the measures and solutions adopted in each system.

Girardi, H. M. and J. K. Grimm. 1979. The history, biology, damage, and control of the gypsy moth. Associated University Presses, Inc., Cranbury, NJ
- "a comprehensive and simplified survey of the history, biology, and control of the gypsy moth"

Hewitt, W. B. and L. Chiarappa, eds. 1977. Plant health and quarantine in international transfer of genetic resources. CRC Press, Cleveland, OH
- a collection of chapters concerning the international transfer of plant pests, parasites, and pathogens. Includes discussion of plant quarantine principles and methods.

Karpati, J. F., C. Y. Schotman, and K. A. Zammarand, eds. 1983. International plant quarantine treatment manual. Food and Agricultural Organization of the United Nations, Rome
- discusses a wide variety of chemical and physical treatment methods known to be useful and effective in plant quarantine situations. Includes treatment schedules and protocols for plant pests associated with miscellaneous materials of plant origin, including garbage.

Laird, M., ed. 1983. Commerce and the spread of pests and disease vectors. Praeger Publ., New York, NY
- based on a 1983 symposium held in New Zealand, this multiauthor publication concentrates on international transfer of medically important pests. It contains considera-

ble discussion regarding pest importation via aircraft.

National Academy of Sciences. 1969. Insect pest management and control. Principles of Plant and Animal Pest Control. Vol. 3. US National Academy of Sciences Publication 1695
- a survey of insect management tactics and approaches that includes an excellent chapter on regulatory control. This text was the source for much of the information presented in this section on regulatory management.

United States Department of Agriculture. 1981. The gypsy moth: research toward integrated pest management. USDA, Washington, D.C.
- a compendium of current understanding and continuing research on the gypsy moth problem in the US. It contains many excellent color photographs.